Attracting
NATIVE
POLLINATORS

Attracting
NATIVE
POLLINATORS

Protecting North America's Bees and Butterflies

Eric Mader
Assistant Pollinator Program Director, The Xerces Society

Matthew Shepherd
Senior Conservation Associate, The Xerces Society

Mace Vaughan
Pollinator Program Director, The Xerces Society

Scott Hoffman Black
Executive Director, The Xerces Society

Gretchen LeBuhn
Associate Professor of Biology, San Francisco State University

Storey Publishing

The mission of Storey Publishing is to serve our customers by
publishing practical information that encourages
personal independence in harmony with the environment.

Edited by Deborah Burns
Art direction and book design by Cynthia N. McFarland
Text production by Liseann Karandisecky and Jennifer Jepson Smith

Front cover photography by © Jim Cane, Mining bee (top left); © Rollin Coville, Metallic sweat
    bee (top center); © Rufus Isaacs, bottom; © Dana Ross, Fender's Blue (top right)
Back cover photography by © William Bouton, Southern dogface; © Rollin Coville, Long-horned
    bee; © Preston Scott Justis, Sachem skipper; © Bruce Newhouse, Mining bee
Spine photography by © Alistair Fraser
Interior photography credits appear on page 373
Illustrations by © David Wysotski, Allure Illustration

Indexed by Samantha Miller

Special thanks to Mars Vilaubi (image management), Ilona Sherratt (illustration coordinator),
Hartley Batchelder (prepress technician), and Elizabeth Stell (botanical editor)

Storey Publishing
210 MASS MoCA Way
North Adams, MA 01247
www.storey.com

Printed in China by R.R. Donnelley
10 9 8 7 6 5 4 3 2 1

LIBRARY OF CONGRESS CATALOGING-IN-PUBLICATION DATA

Attracting native pollinators / by The Xerces Society.
    p. cm.
  Includes index.
  ISBN 978-1-60342-695-4 (pbk. : alk. paper)
  1. Insect pollinators--Conservation. 2. Bees—Conservation.
  3. Insect pollinators--Habitat. 4. Bees--Habitat. I. Xerces Society.
QL467.8.A88 2011
631.5'2—dc22
                                      2010043054

FOREWORD

THIS BOOK IS MUCH MORE THAN A RESOURCE on how to improve habitat for native pollinators. It is a step-by-step guide for changing our stewardship of the earth; it is a tangible way for people of all ages to make a difference. Active participation in this vital, grass-roots revolution is easy: Plant flowers. Sure, by creating floral and nesting habitat, bees, butterflies, and countless other wildlife species will prosper. But through this same simple effort, you will be ensuring an abundance of locally grown, nutritious fruits and vegetables. You will beautify our cities, roadways, and countryside. You will be helping to spread the word about the urgent need to reduce pesticide use, while at the same time creating habitat for beneficial insects that prey upon crop pests. You will be increasing natural diversity and ecological resilience through pollinator gardens, bee pastures, and flowering field borders that stabilize the soil, filter water runoff, and pack carbon into the roots of native prairie plants. For many of our earth's current environmental ills, you will be part of the solution.

— Dr. Marla Spivak
Professor of Apiculture and Social Insects
2010 MacArthur Fellow
University of Minnesota

ACKNOWLEDGMENTS

THE AUTHORS WOULD LIKE TO THANK their families and friends for providing the countless words of encouragement and endless patience throughout long hours of writing and revision. The best parts of this book are a reflection of their support. They also wish to thank Lisa Schonberg for helping develop the bee profiles in the Part 3, Dr. Claire Kremen, Katharina Ullmann, Jessa Guisse, Jennifer Hopwood, Dr. Robbin Thorp, Dr. Neal Williams, Dr. Lora Morandin, Dr. Rachael Winfree, Dr. Marla Spivak, Elaine Evans, Dr. Rufus Isaacs, Dr. David Biddinger, Dr. Jaret Daniels, the many dedicated biologists and conservationists of the USDA Natural Resources Conservation Service, especially Wendell Gilgert, and finally the fantastic staff and individual members of the Xerces Society.

CONTENTS

A New World

FOR EACH OF THE AUTHORS of this book, studying pollinators and creating habitat opened up a new world. We have all been captured by the play that unfolds across a floral stage, waking up one morning to discover male long-horned bees fast asleep in a sunflower, or getting caught up in the intense buzz and energy of a thousand bees in a blooming native basswood tree. We've been surprised to discover bees nesting in unexpected places: wrapped in pink petals under a flowerpot, emerging from tiny holes in a soccer field, or suddenly colonizing the exposed soil in a weeded flower bed. We've been deeply satisfied to watch a leafcutter bee return to a newly mounted bee block with an impossibly big piece of leaf, or a mason bee return with a belly dusted bright orange with pollen. We've stopped and smiled at seeing a flower patch alive with insects.

The delight of this intimate connection with some of nature's most important and hardest workers awaits you. This book will help you get there.

Attracting Native Pollinators is a comprehensive guide to assist gardeners, farmers, educators, park managers, golf course superintendents, and public land managers in understanding, providing, enhancing, and managing habitat for pollinating insects. It focuses on North America, but most of the concepts apply to any of the temperate areas of the world. Using this book as a guide, your efforts on behalf of pollinators will, in turn, help support bountiful harvests on farms and gardens, maintain healthy plant communities in wild lands, provide food for other wildlife, and beautify your landscape with flowers.

Here you will find practical and specific information on how to establish flower-rich foraging patches and nesting sites for pollinating insects any place where there is space enough for a few plants, whether it's a backyard patio in Bangor, Maine, a community garden in New York City, a golf course in the Chicago suburbs, school grounds in Birmingham, Alabama, or a farm in rural California.

This book also includes information on the life history of pollinating insects found in North America, a pictorial guide to native bees, and regional lists of plants that support bees or serve as host plants for butterflies. There are detailed instructions for designing pollinator gardens and habitat, as well as the creation and management of nesting structures for native bees, plus educational activities for children and an extensive list of resources. In short, everything you need to understand pollinating insects — from bees to butterflies, moths, flies, and beetles — and to provide them with food and shelter is here in one book.

Why Focus on Native Bees?

This book devotes considerable attention to bees, especially native bees. We believe this focus is important because of their importance as pollinators and because much of this information has not been widely available to the public. At the same time, scientists learn more each year about the importance of native bees to ecosystems and agriculture, making the story of these vital pollinators even more compelling.

▽ *Bees are the most important single group of pollinators.*

△ *Apples are one of the many crops that require pollinating insects.*

Pollinating wasps have very similar life cycles to bees and share similar habitat needs; conservation measures for bees will assist them, too. Butterflies are frequent flower visitors, but there are already many resources available about butterfly biology and how to create butterfly gardens. Flies and beetles are significant pollinators as well, but while scientists know about their role in pollination, there are no well-established conservation techniques for them. Hummingbirds and some bats are important pollinators for some plants and need flower-rich habitat to sustain them, but as vertebrates, they are beyond the scope of this book. This book stresses bee conservation, but the recommendations provided here will benefit all of these animals.

Though bees are the focus of this book, we have not stressed the European honey bee (*Apis mellifera*), despite its importance as an agricultural pollinator. The honey bee is not native to North America, and there are questions among scientists regarding its impact on native bee populations. Because of its value to agriculture, the honey bee already has the well-organized support of researchers and beekeepers, not to mention a century and a half of books about how to keep them. The goal of this book is to explore the less well-known world of native bees and other pollinators, and to encourage conservation of these animals and protection of their habitat. The practices outlined here for improving flower sources and protecting pollinators from pesticides are the same measures that support habitat for colonies of honey bees.

PART I

Pollinators and Pollination

CUT AN APPLE IN HALF by slicing across its middle and you'll find a central compartment in the shape of a five-pointed star. If the apple has two seeds inside each point of the star — ten all together — it was completely pollinated by bees. If there are fewer than ten, not enough pollen reached the flower's stigmas to develop all of the seeds. A poorly pollinated flower will develop into an apple that's small and lopsided. An unpollinated flower won't develop into an apple at all.

This apple is at the heart of why you should care about pollinator conservation. According to the National Academy of Sciences, close to 75 percent of the flowering plants on the earth rely to some degree on pollinators in order to set seed or fruit. From these plants comes one-third of humankind's food and even greater proportions of the food for much of our wildlife. Yet now, in many places, pollinators are at risk.

▷ *Bumble bees are important pollinators of crops and wild flowers.* Bombus melanopygus *on black raspberry.*

We stand at a crossroad. Honey beekeepers lose unprecedented numbers of their honey bee colonies each year. Once-common bumble bees are disappearing across North America and Europe. Heavily developed agricultural and urban landscapes lack the habitat to support a diversity and abundance of bees, butterflies, and other pollinators.

At the same time, the public no longer overlooks the critical contribution pollinating insects make to the world. Consumers are starting to recognize that the most nutritious and interesting parts of their diet — apples, watermelon, blueberries, carrots, broccoli, almonds, orange juice, coffee, and chocolate, to name but a few — are the result of insect pollination. Wildlife biologists and botanists are teaming up with pollinator conservationists across the world to promote the native plants and habitat management that support this keystone group of animals. Thousands of citizens across North America are noting the bees at their sunflowers for The Great Sunflower Project (see page 212), or collaborating with Monarch Watch to tag monarch butterflies on their way to Mexico.

And each year, researchers learn more about how to manage habitat for wild bees living around our farms and ranches.

It is essential to build on this momentum and to share with friends and family the vital cause of pollinator conservation. By picking up this book, you are joining this effort. You will be prepared to take steps to expand pollinator habitat in your world, whether that be in a small urban backyard, office campus, city park, farm, or natural area.

Beginning Steps

As you think about where to locate your pollinator habitat and how much effort you are prepared to make, consider the range of possibilities. Taking any action, however modest, is better than taking none at all. When it comes right down to it, pollinators have only a few basic habitat requirements: a flower-rich foraging area, suitable host plants or nests where they can lay their eggs, and an environment free of pesticides. Providing any of these is a valuable first step.

Sometimes, simple steps can be the most successful and the most immediately satisfying. Wooden nest blocks or bundles of hollow stems can attract bees within days of installation.

△ *There are many simple ways to connect with native pollinators, from hanging a nest block for mason bees in the garden to creating new greenspaces in urban areas.* ▽

△ *A patch of flowers will offer food to a diversity of insects. Syrphid fly (genus* Eristalis)

A patch of suitable flowers quickly becomes a magnet for butterflies, bees, flies, and beetles. If you use pesticides, seek out alternatives; if your neighbors use pesticides, work with them to find a better way or to minimize their impact.

As you gain confidence in your efforts, increase the scope and size of your pollinator habitat, and you'll add to the diversity, abundance, and productivity of your land. Expand your habitat at your own pace, observe what works and what does not, and incorporate your observations into what you do.

Why Care About Pollinators?

ONSIDER FOR A MOMENT that insect-pollinated fruits and vegetables provide most of the vitamins and minerals we need, and they diversify our diet beyond meat and wind-pollinated grains. The United States and Canada alone grow more than a hundred different crop plants that need pollinators. Without bees, there would be no apples, pumpkins, strawberries, or many other fruits and vegetables.

* Approximately one in three mouthfuls of food and drink require the presence of a pollinator.

* Pollinators provide an ecological service that is essential to the health of the environment.

* Pollinators are central to the lives of other wildlife, from songbirds to grizzly bears.

▷ *Pumpkins are an example of a crop closely associated with specialist pollinators, the native squash bees (*Peponapis *spp. and* Xenoglossa *spp.).*

△ *Bee-pollinated berry crops like these marionberries (above) and blueberries (below) are a mainstay of the Pacific Northwest's agricultural economy.* ▽

The economic value of insect-pollinated crops in the United States in 2003 was estimated to be between $18 and $27 billion. If this calculation is expanded to include indirect products, such as the milk and beef from cattle fed alfalfa, pollinators may be responsible for more than twice this amount. Scientists at the Xerces Society and Cornell University have estimated that approximately 15 percent of this value comes from native bees living in the landscape on and around farms.

In the Pacific Northwest, for example, pollinator-dependent crops are a mainstay of the agricultural economy. In both Oregon and Washington, these crops are concentrated in three agricultural sectors: fruits and berries, alfalfa seed, and vegetable and flower seed. Oregon ranks first in the United States for harvest of blackberries, loganberries, black raspberries, boysenberries, and youngberries. Washington ranks first in the United States for apples, sweet cherries, and pears. Both states also produce substantial crops of vegetable and flower seed that depend on good pollination, and both are major producers of alfalfa seed, whose production requires bee pollination. In 2008, the combined value of pollinator-dependent crops from these two states alone was nearly $3 billion.

Similar examples exist across the country. Maine's 60,000 acres of lowbush blueberries make that state the largest blueberry producer in the United States, with a crop valued at $75 million a year. In New Hampshire and Vermont, the combined 10,000 acres of apple production are valued at more than $30 million a year.

In the Upper Midwest, Wisconsin's cranberry industry contributes $350 million annually to the state's economy and employs more than 7,000 people. With 18,000 acres of cranberry bogs, Wisconsin is the nation's largest cranberry producer. To meet growing export demand, the state is expected to add another 5,000 acres of production within the next decade, all of it requiring bee pollination.

Pollinators also are directly or indirectly responsible for many beverages (such as apple and cranberry juice), fibers (especially cotton and flax), and oil crops such as canola and sunflower, which are increasingly being used for biofuel production. We cannot overlook that pollinators are responsible for many of the beautiful flowering plants and trees in our gardens, which can provide us with spiritual inspiration, better mental health, and improved learning. Research suggests that spending time outside can improve concentration, help the behavior of children with attention disorders, and improve test scores. In short, insect-pollinated plants are the basis for much of our food, our economic stability, and our overall well-being.

BENEFITS TO NATURAL AREAS AND ECOSYSTEMS

The work of pollinators has value beyond providing food and resources for people. Pollinators help keep plant communities healthy and productive. In many areas, pollinators are crucial for flowering plants: for example, almost all of the arid Southwest's bushes and smaller trees — such as mesquite, creosote bush, and chamise — are pollinated by bees.

▽ *For long-lived plants such as kinnikinnick (Arctostaphylos uva-ursi), a loss of pollinators will not be noticed for many years.*

When the pollinators are lost, bushes and trees may continue to flower and look normal for decades. By the time someone notices that they are not reproducing, it may well be too late to reintroduce a pollinator and preserve the ecosystem.

In the case of some rare species, only by identifying and understanding the

△ *Mining bee (*Andrena *sp.) nest entrances are easily overlooked, but represent an essential ecological keystone.*

habitat needs of their pollinators can we even begin to ensure the plants' survival. Protecting the plant's habitat isn't sufficient. The endangered dwarf bear poppy (*Arctomecon humilis*), for example, grows only in the Virgin River basin of southern Utah. The flower's dedicated pollinator — the minuscule Mojave poppy bee (*Perdita meconis*) — was finally identified nearly a decade after the plant was listed under the Endangered Species Act. Only then could conservation efforts focus on all aspects essential for its survival, including providing nesting sites for its pollinator.

Supporting Plants and Animals

Pollinators are not only essential to plant reproduction, but they help plants in other ways as well. The larvae of some flower-visiting beetles bore holes in dead or weakened trees, thereby initiating decay and returning the nutrients locked away in the tree back into the soil to be used by other plants.

In some areas, pollinators support plant communities that in turn stabilize the soil, thus preventing erosion and helping to keep creeks clean for fish, mussels, and other aquatic life. Willow shrubs and trees, for example, are important for the stabilization of stream and river banks. Because there are separate male and female plants, insects transfer much of the pollen from the male to the female willow flowers. Goldenrods, another example, are important colonizers of disturbed habitat. They too depend upon insects to produce their abundant seed.

Mammals, ranging from red-backed voles to grizzly bears, depend on insect-pollinated fruits and seeds. In the late summer, fruits can account for more than 60 percent of a grizzly bear's diet. About a quarter of all birds consume, as a major

▽▷ *Insect-pollinated seeds and fruits feed countless other wildlife species, from grizzly bears to cedar waxwings.*

△ *Pollinators are an integral element of terrestrial plant communities.*

part of their diets, the fruits and seeds that result from animal pollination. Pollinating insects themselves are food for birds, lizards, and spiders, and as such are a central part of the food web. More than 90 percent of birds rely on insects during at least one stage in their life. Studies have demonstrated that diverse plant communities along stream banks, such as those that support pollinators, also support more diverse and abundant communities of insects that fall into the adjacent streams and become food for fish.

Managing Habitat for Pollinators

Considering the central role that pollinating insects and their habitat play in functioning ecosystems, it's easy to see how pollinator conservation can provide a framework for habitat management that helps other plants and animals. Until recently, however, habitat management has not considered pollinator habitat. In the Midwest, for example, restored prairie is maintained with burning or mowing. This site maintenance traditionally has focused squarely on the plant communities, with little thought to impacts on pollinators such as bees and butterflies. Supporting pollinators, on the other hand, requires limiting mowing or burning to a third or less of the site at any one time so that there is no year where all floral resources are unavailable. This results in more of a mosaic environment that allows for colonization of disturbed sites from the nontreated areas and a more diverse habitat.

BENEFITS TO PEST MANAGEMENT

Many insects that visit flowers as adults to feed on nectar or pollen also provide pest control for plants. For example, the larvae of many syrphid flies eat aphids. The adult fly fuels itself by eating sugary nectar. It then searches for plants with aphids where it lays eggs. The maggot that hatches from the egg patrols the plant looking for aphids, which it grabs onto and sucks dry, leaving behind a trail of shriveled bodies. An abundance of small flowers, such as those found on yarrow, fennel, or alyssum, increases the number of eggs an adult syrphid can lay over her lifetime. So, while the fly may provide some pollination, it also provides plants with protection from the pests that would harm them.

Still other pollinators are parasitoids that lay their eggs inside plant-eating aphids and caterpillars. Some of the most common parasitoids are tachinid flies and braconid and ichnuemonid wasps. When parasitoid eggs hatch, the larvae slowly feed on their host. During this period, the host typically is less active than it would be otherwise, doing less damage to the plant where it is feeding. Then, when the parasitoid larvae have completed their feeding and development, they kill their host, pupate, and emerge as adults.

△ As adults, tachinid flies are pollinators. As larvae, they typically live as internal parasites of other insects.

CONSERVATION

Given the importance of pollinators, scientists have been documenting with particular alarm the dramatic declines in populations of bees and other pollinating insects in recent decades.

△ Two species of bumble bees are giving scientists particular cause for concern: yellow-banded and western.

In the mid-1990s, the yellow-banded bumble bee (*Bombus terricola*) was the most abundant bumble bee in Wisconsin. Ten years later, it made up less than 1 percent of the state's bumble bees. Across the continent, a similar fate has befallen the western bumble bee (*B. occidentalis*). Once one of the most abundant bumble bees on the West Coast, its numbers have crashed since the early 1990s and it is rarely seen now.

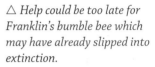

△ *Help could be too late for Franklin's bumble bee which may have already slipped into extinction.*

It gets worse. Franklin's bumble bee (*B. franklini*), restricted to a small area along the Oregon-California border, may have gone extinct during the same period. The most likely cause of these sudden declines is the spread of disease from commercially reared bumble bee colonies.

"Oh, a few bees," you may think. "Why worry?" Bees are the preeminent pollinators in North America. Pollinators are a keystone species group; the survival of a large number of other species depends upon them. As mentioned, they are essential to the reproductive cycles of most flowering plants, supporting plant populations that animals and birds rely on for food and shelter. Pollinators are also indicator species, meaning that the viability and health of pollinator populations provide a snapshot of the health of the ecosystem. As the insects that many plants require for adequate pollination disappear, the effect on the health and viability of crops and native plant communities can be disastrous.

In California, the endangered Antioch Dunes evening primrose (*Oenothera deltoides* ssp. *howellii*) survives on only a few acres of degraded sand dunes; it lacks pollinators and thus produces only a fraction of its potential seed crop. The rare white fringed orchid (*Platanthera praeclara*), scattered across grasslands of the midwestern United States, is now visited by a species of moth that is not native to the habitat. This moth fails to pollinate the flowers because its long tongue can reach the nectar without its head and body touching the pollen. Furthermore, it drinks nectar that might otherwise attract hawk

△ Tens of thousands of honey bee hives are required to pollinate Maine's blueberry fields.

moths with shorter tongues, which are the orchid's legitimate pollinators.

Each spring in Maine, more than 60,000 honey bee hives are unloaded from semitrailers into the state's expansive low-bush blueberry farms. Each hive contains more than 20,000 bees that have made the overnight journey from Florida or the Carolinas. They are there, in the blueberries, because modern industrial-scale agriculture has reduced the area of habitat available to support the more than 270 bee species that are native to the state. And when the blueberry bloom is over, the lack of forage means that the bees will be moved to other crops in other states. The native bees are in fact far superior pollinators of blueberry than the nonnative honey bee, having evolved with the wild blueberry plants in Maine's pine barrens over thousands of years. What these native bees lack is habitat: food available from spring to fall.

In China's Sichuan Province, one of the largest apple-producing regions in the world, farmers perch on ladders in mountainside orchards to pollinate the blossoms by hand. The farmers have adopted this practice because wild bees are now absent in their area, and honey beekeepers refuse to bring in their hives due to excessive pesticide use in the orchards.

THE RED LIST
OUR ENDANGERED POLLINATORS

THE FACT THAT POLLINATING INSECTS face threats and are in need of conservation assistance is well established. Too often, however, a lack of information about particular species hampers efforts to focus research funding or conservation action onto those in greatest need. Important information that can guide the decisions of policy makers and land managers is often hidden in scientists' files, or is lacking altogether.

To overcome this roadblock, the Xerces Society for Invertebrate Conservation developed the Red List of Pollinator Insects of North America. The objectives of the list were fourfold:

* To identify those species of pollinating insects at greatest risk;
* To compile information about them;
* To identify gaps in knowledge that could lead to new research;
* To use this information to launch protection efforts.

The Red List includes 59 butterflies and moths and 57 bees. A detailed profile for each pollinator identifies its habitat needs and distribution; gives information on biology, status, and conservation needs; and lists critical threats to its survival. The information was gathered from an extensive review of published literature and gleaned from numerous researchers. In addition, Dr. John Ascher, American Museum of Natural History; Dr. Karl Magnacca, University of Hawaii – Hilo; Dr. Laurence Packer, University of York; and Dr. Robbin Thorp, University of California – Davis each wrote profiles. The project was supported by a grant from the CS Fund.

Since the Red List's release in May 2005, it has had a significant impact. In 2006, the National Research Council committee on the status of pollinators in North America requested a presentation about this summary of imperiled pollinators, and the Red List was subsequently reprinted as an appendix in the committee's report, *Status of Pollinators in North America*. It has guided and inspired research projects of numerous scientists, and conservationists have used the information gathered into the profiles to launch efforts for protecting of pollinators and their habitat. Specifically, Xerces has asked that several species of highly imperiled yellow-faced bees (*Hylaeus*) in Hawaii be protected under the Endangered Species Act.

The Red List profiles also provided the foundation for the Xerces Society's multiyear initiative to protect bumble bees. Focusing on the rusty patched, yellowbanded, and western bumble bees, this project enlisted the aid of volunteers to search for additional populations and recruited scientists from across the United States to help complete comprehensive status reviews. One of the principal threats identified was disease spread through the commercial rearing and shipping of bumble bee colonies. Supported by more than 50 scientists, the Society subsequently asked the U.S. Department of Agriculture to develop regulations to control interstate movement of bumble bees.

It is hoped that the Red List will continue to allow scientists to identify knowledge gaps and prioritize research needs, and that it will be used to support additional advocacy for those pollinators in greatest need. The Xerces Society Red List of Bees, and Red List of Butterflies and Moths, are available online at *www.xerces.org/red-lists*. ❀

Finding Solutions

Four factors — the loss and fragmentation of habitat, the degradation of remaining habitat, pesticide poisoning, and the spread of diseases and parasites — account for most of the declines in populations of bees and other pollinators. These factors have complex political, economic, and social origins that are not easily addressed. At the local level, however, the solutions to many of these problems are simple and straightforward. Many insects are fairly resilient, and there are actions we can take in our own backyards and neighborhoods, on farms and ranches, and in city parks and wild areas, to help strengthen and support pollinator populations.

The imperative behind this book is the fact that, as with most wild creatures, pollinator populations are declining as a result of human activity. If this trend continues, driving species of pollinators to extinction, the result will be disastrous not only for the insects but for humans as well. This book provides tools — information, understanding, and practical suggestions — that will help you begin to reverse this trend by providing and enhancing pollinator habitat and by becoming more aware of how your actions affect pollinators.

The Biology of Pollination 2

POLLINATION IS CENTRAL to the life cycle of flowering plants. Pollen must move from male to female parts of flowers for the plant to develop seed and reproduce. Movement of pollen within a flower or between flowers on the same plant is called **self-pollination**; the movement of pollen among flowers on different plants is **cross-pollination**. Although some plants can produce seed from self-pollination, most require cross-pollination; many plant species actually have genes that prevent their own pollen from fertilizing their ovules.

* Pollination is the process by which plants reproduce.

* Nearly 75 percent of all plants on earth require animals for pollination.

* Plants and their pollinators evolve together, each constantly adapting to the other.

△ *Bees and other pollinating insects are essential for moving pollen among flowers.*

In addition, a transfer and mixing of genes occurs when pollen moves among plants. Over multiple generations, this results in more vigorous plants that constantly evolve to thrive in changing conditions.

How Pollination Works

Understanding pollination starts with understanding the roles of each different part of a flower. Like all organisms that rely on sexual reproduction, flowering plants have both male and female structures. Pollen grains — vessels for a plant's male **gametes** (sex cells) — need to move from the former to the latter to achieve fertilization. The male structure, called the **androecium**, is made up of the **stamen**, which consists of a bulbous **anther** on the end of a thin **filament**. The anther produces pollen. The female parts are collectively known as the **pistil**, formed of one or more **carpels**. Carpels contain the female gametes. Each carpel has three principal parts: a sticky or feathery tip, called the **stigma**; an **ovary** buried in the base of the carpel; and between the two, a stalklike extension called the **style**.

In some flowers, such as those of wild roses, apple trees, or cacti, the stamens and carpels can usually be clearly seen. The

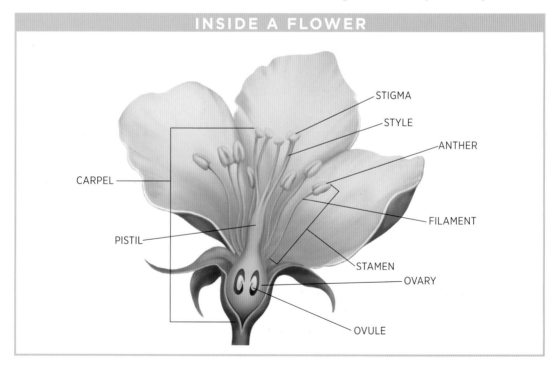

INSIDE A FLOWER

STIGMA

STYLE

ANTHER

CARPEL

FILAMENT

PISTIL

STAMEN

OVARY

OVULE

stamens form a ring at the base of the petals, surrounding the central pistil. In other plants, such as tomatoes, the stamens are fused together and surround the pistil.

The number of both stamens and carpels in a flower ranges from none to many. When the male and female parts occur together in the same flower, it is referred to

△ *The flower structures of a wild rose are relatively easy to identify.*

as a **complete** or **hermaphroditic** flower. When a flower is lacking either stamens or carpels, it is called an **incomplete** flower, and will be male or female. Male and female flowers may also occur on separate male and female plants in the case of **dioecious** (Latin for "two houses") species, or both may be found on the same plant in the case of **monoecious** (Latin for "one house") species. To round out this complex mixing and matching of male, female, and combined flowers, there also are rare cases of **androdioecious** and **gynodioecious** plants where one type of plant has hermaphroditic flowers and the other type of plant has either male-only (**androdioecious**) or female-only (**gynodioecious**) flowers.

Flowers may occur singularly, as **simple** flowers such as lilies, or in complex clusters of flowers called an **inflorescence**, such as the spike of flowers on the lupines. Sunflowers and other daisies are an extreme example of an inflorescence; what appears as a single flower head is, in fact, hundreds of tiny individual flowers.

Germination and Fertilization

△ *The head of a sunflower is made up of hundreds of individual flowers. Each one needs a bee visit in order to produce a seed.*

Whatever the configuration, when a pollen grain reaches the stigma, it germinates and sends a microscopic tube down inside the style, carrying the male gametes to the plant's ovules within the ovary, resulting in fertilization. Each fertilized ovule then becomes a hard seed from which a new flowering plant will grow. When pollen is not transferred, no seed is formed. For flowers such as apples or watermelon that contain multiple ovules, if only a small amount of pollen is transferred between blooms, not all of the ovules may become fertilized. The result is fewer seeds and smaller or lopsided fruit.

The pollen that lands on a stigma can be **self-pollen**, which means it came from the same plant, or **outcross pollen**, meaning it came from a different plant. Many plants have developed traits that minimize self-fertilization. For example, there are genes that prevent the growth of pollen in a stigma that is genetically similar. Other plants do not produce pollen and ovules at the same time. Dioecious plant species actually have different male and female plants so there is no possibility of self-pollination. About 50 percent of all flowering plants are able to develop seeds from self-pollen; however, the resulting seeds and offspring are often not as vigorous as those from outcross pollen.

HOW POLLEN TRAVELS

Since flowering plants are literally rooted to one spot, they have evolved ways to spread their pollen, increasing the probability of outcrossing. About half of all flowering plants require pollen that comes from another plant.

The wind carries the pollen of some plants, including grasses, conifers, and ragweed. The likelihood of wind-borne pollen reaching a flower is low, so the blossoms of these plants must produce millions of pollen grains to be carried on the breeze (and cause misery for hay fever sufferers) to reach even a few receptive targets.

The majority of flowering plants — close to 75 percent — rely upon animal pollinators to move pollen. Because animals transfer pollen more efficiently than wind and water, such plants are able to produce pollen in smaller quantities. These grains of pollen are larger and have rough surfaces that help them cling to pollinators. The pollen of bee-pollinated flowers also has a higher nutritional value than the pollen of wind-pollinated plants. Its protein level may be as high as 60 percent, and it also may include other important nutrients such as vitamins, minerals, lipids, and sterols.

ATTRACTING VISITORS

To attract pollinators, plants have developed flowers to advertise their presence. Brilliantly colored petals stand out against a backdrop of green foliage and serve as a landing platform where insects rest while feeding. Perfumed smells carried on

△ *Grasses spread pollen on the wind.*

VOYAGING BY WATER

In some plants, male gametes float or are splashed from one plant to another by water. With mosses, for example, pollen may travel in a splash of rain; with certain aquatic plants, it may ride on the surface of a pond or stream.

wind currents attract distant insect visitors. To reward their pollinators, plants provide resources such as nectar (a sugary secretion) or pollen (a protein source).

The shape and color patterns of flowers often provide passing insects with additional cues to help them find a flower's nectar. For example, many flowers have patches of color forming radiating lines

△ *The shape or markings on flowers help to guide insects to the rewards of nectar or pollen.*

or targetlike concentric circles that direct insects to the nectar. In many flowers, these **nectar guides** contain ultraviolet color patterns that bees can see but people can't.

Once they are fertilized, flowers stop producing nectar, as they no longer need to attract pollinators. They may also change color, an obvious visual signal to insects that there's no more food and that they should attend to other, pollen-laden flowers.

Pollination Syndromes

The collective traits of different flowers — including their shape, size, color, odor, and time of bloom — are sometimes referred to by botanists as **pollination syndromes**. The type of pollination syndrome that a particular plant exhibits suggests which animals are most likely to visit its flowers. Red flowers seem to attract animals that can see the color red, such as hummingbirds, whereas blue flowers are more likely to attract insects that see in the ultraviolet end of the spectrum, such

△ *Leafcutter bee (genus* Megachile) *on ceanothus*

△ *Tiger moth* (Gnophaela vermiculata)

△ *Boisduval's blue* (Plebejus icarioides)

as bees. Birds are unable to smell, so most hummingbird-pollinated flowers do not have a scent.

The theory of pollination syndromes has helped scientists understand the process of **convergent evolution**, whereby plant and pollinator interactions lead to increasingly specialized adaptations on the part of each. This specialization can be extremely

△ *Red flowers typically attract hummingbirds or butterflies.*

rewarding. Pollen and nectar take a lot of energy to produce. The energy invested is worth it only if that plant can then attract more dependable and efficient pollinators, improving its chances of reproductive success. For the pollinator, foraging is similarly an energy-intensive activity and exposes it to predation and bad weather. If the flowers match its body type and foraging preferences, however, visiting them requires less work for the pollinator to earn the same reward and offers greater reproductive success.

Extreme specialization on the part of either plant or pollinator is risky, as the disappearance of one can lead to the disappearance of the other. Consequently, extreme specialization is rare. And pollination syndromes are not necessarily accurate predictors of plant-pollinator interactions. One study, for example, suggests that on average only about one-third of plant species can be accurately classified according to pollination syndromes. Therefore, although pollination syndromes are a starting place to identify which plants might support particular pollinators, other plant selection criteria, such as those outlined later in this book, should always be used.

△ *Digger bee* (Anthophora urbana)

Most animal pollinators need a diversity of flowers in order to survive, and most plants need a diversity of pollinators. Together both plant and pollinator constantly adapt to seasonal change, predation, disease, and changing environments in a complex dance of interdependence. Their individual survival depends upon one another, and ours upon them.

Meet the Pollinators 3

POLLINATORS DON'T GATHER pollen with the intent of promoting plant reproduction. They are animals looking for food, and they've found the sweet nectar and fatty, protein-rich pollen to be excellent foods for themselves or their young. They may also be looking for mates, sometimes meeting at flowers where they gather food, or in other cases collecting floral oils to use as perfumes to attract a mate. As pollinators feed, court, or gather oils, pollen grains stick to their bodies and rub off accidentally when they visit other blossoms.

❀ While bats, birds, and even lemurs can pollinate flowers, the vast majority of pollinating animals worldwide are insects.

❀ There are four major groups of pollinating insects: bees and wasps, flies, butterflies and moths, and beetles.

❀ While important to agriculture, the honey bee is not native to North America.

△ *The hairiness of bees is one reason they are such effective pollinators.*

Across the globe, pollinators range in size from such minuscule insects as the fig wasp (only 1.5 mm long) to 10-pound Madagascan lemurs. While the number of pollinator species is unknown — estimates vary from 130,000 to 300,000 — by far the greatest number are insects.

Comparing Insect Pollinators

Bees are the most important group of pollinators. With the exception of a few species of wasps, only bees deliberately gather pollen to bring back to their nests for their offspring. Bees also exhibit a behavior called **flower constancy**, meaning that they repeatedly visit one particular plant species on any given foraging trip. This is important because pollen is wasted if it is delivered to the wrong species of flower.

On a single foraging trip, a female bee may visit hundreds of flowers, transferring pollen the entire way. In contrast, butterflies, moths, flies, wasps, and beetles visit flowers to feed on nectar (or on the flower itself, in the case of some beetles) and not to collect pollen. Thus, they come in contact less frequently with the flower's anthers than bees do.

INSECT LIFE STAGES

All of the pollinating insects described in this book undergo the same basic life cycle: complete metamorphosis through four distinctly different life stages from egg to larva to pupa to winged adult. Pollinating insects vary considerably, though, in where the female lays her eggs, how mobile the larvae are, and what they eat during their various life stages.

The life span from egg to death may last from a few months to a couple of years. Along the way, the insect's shelter and food requirements will change to match the different stages of its life.

Butterflies lay their eggs directly on or near the caterpillar's food source, which is often a particular species of plant (the **host plant**) so that the caterpillars don't have to crawl far for nourishment. Species of flies and beetles lay their eggs similarly close to (or on) the larval food source, which might be rotting wood, aphids, or soil-dwelling invertebrates. Bees and many wasps, in contrast, lay their eggs in protected brood cells within a nest. The larvae are not mobile, so the mother must provision these natal cells with enough food to nourish the brood.

After the larva hatches from the egg, it feeds upon its preferred food source and will take two weeks to more than a year to grow through its larval stages. Most bees and wasps complete larval development in four weeks or less. They have all the food they need on hand, provided by the mother or — in a small number of cases — by workers in a colony. They also have generally stable conditions in their nests. Flies, butterflies, and beetles are much more variable: the duration of the

△ *Freshly hatched caterpillars crawl away from the egg cluster.*

△ *Caterpillars need the right plants on which to feed.*

larval stage depends upon weather conditions, palatability and abundance of their host plants or insects, and the time of year the egg is laid.

Pupation

After they have eaten and grown through several molts, the larvae pupate and drastically change their appearance and behavior from a larva, caterpillar, grub, or maggot to a winged adult. Pupation takes many forms, depending upon the species.

Within a nest. In the case of bees and many wasps, the larva defecates, weaves a silken cocoon around itself, and then pupates in its cell within the nest.

In or on a host. Parasitoid wasps and flies — the larvae of which grow inside the body of a living host insect — will defecate (which usually kills the host) and then pupate either within or on the surface of the host.

At a distance from a host plant. Butterflies and moths may demonstrate a roaming behavior during their final days as a caterpillar. In this case, they leave their host plant and crawl about, often in a seemingly random direction. This gets them far from where they were eating and helps them avoid wasps or other predators and parasitoids that may have been attracted to the scents emitted by their feeding. The larva eventually finds an appropriate place to form a cocoon or chrysalis and then pupates.

Near a food source. Beetles and nonparasitoid flies vary in where they pupate, either staying within their food source or dropping to the ground and pupating in the litter on top of the soil.

Adult Activity

Pollinators vary significantly in how long they are active as adults, but for the majority it is a matter of a few days or weeks. Most individual bees remain active as adults for three to six weeks. Individual bumble bee workers may live longer, typically four to six weeks, and the bumble bee queen will live for a full year. Honey bee queens live even longer. Adults of a particular species, however, may emerge from their natal nests over the course of a few weeks, resulting in a longer active adult period for the species.

△ *The pupal stage of a butterfly — the chrysalis — may happen on the host plant or nearby.*

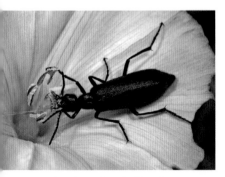

△ *Blister beetle (family Meloidae)*

Flies and beetles have short adult lives, ranging from a few hours to several weeks. Butterflies live as adults more briefly, usually only a week or two. Inevitably, there are exceptions. A few butterfly species, such as the monarch, California tortoiseshell, and the mourning cloak, overwinter as adults and thus may live for several months.

Knowing the life span and habitat needs of each life stage is important to accommodate pollinators. To supply habitat for a variety of species, you must provide a diversity of plants, nesting sites, and hibernation habitats to match their changing life stages throughout the year. To simplify creating and managing favorable pollinator habitat, this book focuses on the life history needs of the most important pollinators in temperate North America: bees. If you meet the forage, shelter, and overwintering needs of a diverse community of native bees, you will create habitat for other pollinators as well.

BEES AS POLLINATORS

Bees are considered the most important group of pollinators for a simple reason. Female bees collect nectar and pollen from flowers as food for their offspring and, in doing this, accidentally transfer large quantities of pollen from flower to

WHAT DISTINGUISHES A BEE?

ENTOMOLOGISTS distinguish bees from other insects by the constricted waist between thorax and abdomen, two pairs of wings, their mouthparts, and segmented antennae that are long and cylindrical. Seen through

scopa

a microscope, individual body hairs appear branched or feathered, which differentiates them from their closest relatives, the wasps, who have hairs that are not branched. In addition, many female bees have evolved structures for transporting pollen: either a patch of long, stiff hairs (called a **scopa**) on each hind leg or under the abdomen for carrying pollen, or a pollen basket (a **corbicula**), a space fringed by long, inward curving hairs on each of their hind legs for pollen moistened with nectar.

△ *North America's smallest bee*, Perdita minima, *laid on the face of its largest*, Xylocopa varipuncta.

flower. Both male and female bees feed on nectar, but only the females gather pollen to take back to their nests.

Bees are immensely diverse insects that form an important group within the Hymenoptera, an insect order that also includes ants, wasps, and sawflies. Worldwide, there are an estimated 20,000 species of bees, with approximately 4,000 species native to the United States. North American species range in length from about 1/12 inch to more than 1 inch (2–25 mm). They vary in color from dark brown or black to red or metallic green and blue; some have stripes of white, orange, yellow, or black; and some even have opalescent bands.

NATIVE BEES COMPARED TO HONEY BEES

THE MAIN ADVANTAGE of honey bees as crop pollinators is that they can be supplied in large numbers and are readily transportable. But how do they compare at the actual job of crop pollination?

In some cases, native bees pollinate apples, cherries, squash, watermelon, blueberries, cranberries, and tomatoes far more effectively than do honey bees on a bee-per-bee basis. Many native bee species also forage over a longer period of time — earlier or later in the day — than honey bees. Finally, native bees will often visit flowers in wet or cold conditions when honey bees remain in the hive.

In North Ogden, Utah, when blue orchard bees (*Osmia lignaria*) were used to pollinate cherry orchards, the average fruit production was more than double that of when honey bees were used. In addition, in bad weather years, when neighboring orchards did not harvest, the blue orchard bee–pollinated orchard had a harvestable crop. The blue orchard bees were very efficient in the cherry orchard for several reasons. They have a short foraging range, which means they rarely leave the orchard. In addition, the way they manipulate the flower means they make contact with anther and stigma on almost every visit. Finally, blue orchard bees were active at light levels and temperatures that were too low for honey bees to forage. The result was that over a period of five days, blue orchard bees spent more than twice as long foraging — 33 hours compared to 15 hours by honey bees.

Native bees can make honey bees more effective as pollinators by causing them to move more frequently between rows of male and female plants. Plant breeders at one hybrid sunflower seed farm in California choose to plant varieties that are particularly difficult to cross-pollinate with each other in areas of their fields adjacent to riparian habitats that support native bees. The breeders find that these fields have greater numbers of native bees and the sunflowers produce correspondingly more seeds.

Their common names include plasterer bees, leafcutter bees, mason bees, carder bees, digger bees, and carpenter bees, reflecting the many ways they build nests. Others are named after a particular behavior. Sweat bees lap up perspiration for the salt; bumble bees hum loudly as they fly. Cuckoo bees lay their eggs in the nests of other bee species, and honey bees make and store honey.

A bee undergoes the same four stages of metamorphosis as other insect pollinators described in this book. Only the last of these life stages, the adult, is recognizable as a bee. During the first three stages, the bee is inside the brood cell of the nest. How long each stage lasts varies widely by species.

△ *Common names of bees often reflect lifestyle traits. This cuckoo bee will lay its eggs in the nests of other bees.*

Types of Bees

Bees can be categorized in several ways: whether they are solitary or social species, whether they nest in the ground or in a cavity, whether they build and provision their own nests or parasitize the nests of others, and by their foraging habits. **Generalists** are bees that gather nectar and pollen from a wide range of flower types and species. The majority of bees are generalists. **Specialists**, on the other hand, tend to use a single plant family or genus for their pollen needs. The life cycle of specialist bees is often closely tied to their host plants; the new generation of adults emerges from their brood cells when the host plants are flowering.

SOLITARY BEES

Of the roughly 4,000 species of bees in North America, more than 90 percent lead solitary rather than social lives, each female constructing and provisioning her own nest without any help from other members of her species.

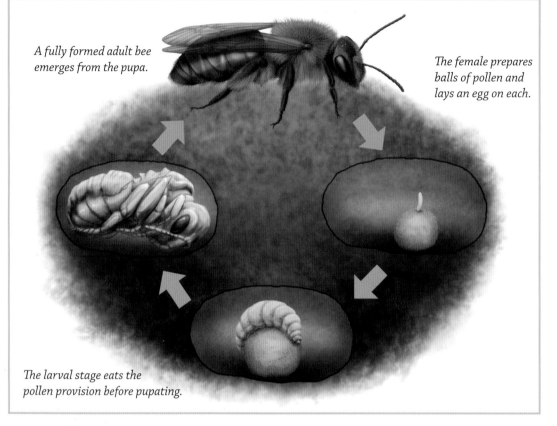

A fully formed adult bee emerges from the pupa.

The female prepares balls of pollen and lays an egg on each.

The larval stage eats the pollen provision before pupating.

Life Cycle

Solitary bees usually live for about a year, although humans only see the active adult stage, which lasts about three to six weeks. These insects spend their early months hidden in a nest, growing through the egg, larval, and pupal stages.

EMERGENCE

When a male bee emerges from the nest, he typically loiters around a nesting area or finds a patch of flowers where females might be foraging, hoping to mate. After a female bee emerges, she mates and then spends her time creating and provisioning a nest in which to lay eggs. After a few weeks, during which she may have prepared and provisioned two dozen or more brood cells, she dies.

Typically, the adults of a species emerge at roughly the same time each year: for example, early spring in the case of blue orchard bees (*Osmia lignaria*) or midsummer in the case of squash bees (*Peponapis* species). A few solitary bees, such as

△ *Squash bees (*Peponapis pruinosa*) in a squash blossom*

some sweat bees in the genera *Halictus* and *Lasioglossum*, have two or three generations each year and so are present over a long period of time.

NESTING

The many solitary bee species exhibit a range of nesting behaviors. Some share a nesting site, sometimes in large aggregations or at great densities. For example, in well-established nesting sites of alkali bees (*Nomia melanderi*), 200 or more females may nest in a single square yard of ground, each bee tending a separate burrow. A few "solitary" species are communal in that they share a common entrance tunnel to the nest, but each female creates her own brood cells. More unusual are the large carpenter bees (genus *Xylocopa*): these females live long enough to meet — and share the nest with — their adult offspring.

Female solitary bees have amazing engineering skills, going to extraordinary lengths to construct a secure nest. About 30 percent of solitary bee species use abandoned beetle burrows or other tunnels in **snags** (dead or dying standing trees). Alternatively, they may chew out a nest within the soft central pith of stems and twigs. The other roughly 70 percent nest in the ground, digging tunnels in bare or sparsely vegetated, well-drained soil. A few species nest in eclectic places such as empty snail shells and potlike cells that they construct on twigs from pebbles and tree resin.

Each bee nest usually has several separate brood cells in which the female lays her eggs, one egg per cell. The number of cells varies by species. While some nests may have only a single cell, most have five or even many more. Female wood-nesting bees make cells in a single line that fills the tunnel. Females of ground-nesting species may dig complex, branching tunnels. To protect the developing bee and its food supply (from drying out, excess moisture, fungi, and disease) the cell may be lined with waxy or cellophane-like glandular secretions, pieces of leaf or petal, mud, or chewed-up wood.

△ *Leafcutter bees (genus* Megachile) *use leaf pieces to partition brood cells and seal off nest entrances.*

LAYING

Before she closes each cell, the bee provisions it with food for her offspring. She mixes together nectar and pollen to form a loaf of "bee bread" inside the brood cell. She then lays an egg in the cell, usually on the loaf, and seals the cell. When she has completed and sealed each of the cells in the nest, the bee will close off the nest entrance and leave.

A female solitary bee may lay up to 20 or 30 eggs in her life. Each egg resembles a tiny white sausage. As she lays each egg, the female can control whether it will hatch into another female or a male. She can do this because after mating she stores sperm in a special sac called a **spermatheca**, releasing it only when needed during egg laying: female eggs are fertilized, male eggs are not.

LARVA TO PUPA

One to three weeks after it is laid, the egg hatches and a white, soft-bodied, grublike larva emerges. All of the bee's growth occurs during this larval stage. Feeding on the bee bread, the larva grows rapidly for several weeks, passing through four or five larval stages of development (**instars**) before changing into a pupa. To avoid contamination of the larva's food supply, the connection between the midgut and hindgut that allows defecation does not develop until the final instar. The last part of the final instar's development, after it has finished feeding and has defecated, is called the **prepupa**.

The length of time that bees remain in the prepupal stage varies. Solitary bees may have a year or more between generations and can remain dormant as prepupae for months during winter or periods of drought or other unfavorable conditions. In general, solitary species that emerge in spring (the

△ *Metallic sweat bee* (Agapostemon texana)

PERFECT TIMING

THE CHANGES from prepupa to pupa and then to adult — and the adult's subsequent emergence — are most likely triggered by a variety of factors, such as the internal clocks of the bees or environmental cues like moisture level and temperature, so that the adults will be active when the flowers they need for foraging are in bloom. Some bees nest deep underground — 18 inches or more — where the moisture and temperature of the soil do not fluctuate noticeably. The southeastern blueberry bee (*Habropoda laboriosa*) is one such example: although it is not known what cues trigger the bees' emergence from these stable nest conditions, they emerge just as the blueberry plants they use for forage begin to flower.

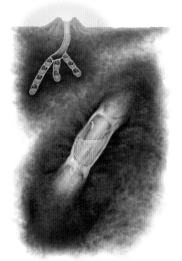

Mining bees (genus Andrena) excavate multiple underground brood cells that usually radiate outward from a single vertical entrance shaft.

Colletes *bees typically construct underground rows of brood cells protected by a waterproof, plasticlike, glandular secretion.*

Megachile perihirta, *one of the few ground-nesting leafcutter bees, protects brood cells in rolled layers of leaf pieces or flower petals.*

*Most mason bees (*Osmia *spp.) nest in hollow stems or holes bored into wood. They provision a linear series of cells with pollen, separating each with a mud or leaf wall to protect the brood from predators.*

Ground-nesting bees sometimes excavate soil around the hole just as ants do.

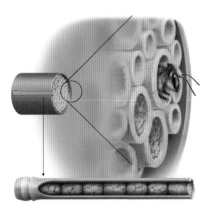

Leafcutter and mason bees may nest in a bucket or coffee can filled with sections of hollow bamboo stems.

blue orchard bee, for example) complete pupation by the end of the summer and overwinter as fully formed but dormant adults. In arid regions, where the emergence of adult bees is synchronized with the infrequent blooming seasons of desert flowers, the prepupal stage of some species will last for a year or more, waiting for rain to spur the plants into flowering. One researcher was very surprised to have bees emerge in his office from desert soil he had collected more than 10 years earlier.

Foraging Habits of Solitary Bees

When foraging for food, most bees search for two things: nectar for energy, and pollen to feed their brood. Some bees also collect special oils from flowers that are used as both an adult and larval food source. Depending upon the size of the bee, it may travel from 100 feet to more than a mile from its nest.

Solitary bees usually collect nectar from any flower structure that allows them to reach the sweet liquid. The range of flowers from which bees can gather nectar depends upon the length of their tongues. Short-tongued bees can drink only from open flowers such as asters or daisies; long-tongued bees can reach the nectar offered by deep or complex flowers such as penstemon, lobelia, and lupines.

HOW SELECTIVE ARE BEES?

BEE SPECIES GATHER NECTAR from a variety of sources but vary in how particular they are about where they collect pollen. In general, solitary bee species are **polylectic** — floral generalists that will gather pollen from many kinds of flowering plants. Because their diet is so varied, polylectic bees may be relatively adaptable to changes in plant communities and degraded habitats.

Some species are more selective. **Oligolectic** bees collect pollen from a small group of closely related plants, perhaps a single plant family or genus. The aptly named squash bees (genus *Peponapis*) specialize on the pollen of squash and pumpkins. The seven U.S. species in the megachilid (leafcutter) subgenus *Lithurgopsis* gather pollen only from cacti. *Diadasia* is an interesting bee genus because different species are oligolectic on different plant families: in the Southwest, species of *Diadasia* specialize on cacti and mallows, while *Diadasia* species elsewhere specialize on evening primroses or sunflowers.

Monolectic bees are more specialized still; they gather pollen from just a single species of flower. Because of the restricted range of their food sources, monolectic bees are particularly vulnerable to changes in their habitat.

ANOTHER SIGNIFICANT INFLUENCE on foraging is how far the bees can fly. For a bee to provision a nest successfully — or even construct it, if nesting materials are required — the nest site and forage area must be within flight distance. How far a bee can fly depends on its size.

* A large bee can fly farther than a small one. Large carpenter bees (*Xylocopa*) and bumble bees (*Bombus*) can forage at distances of a mile or more from the nest.
* Medium-sized bees such as mining bees (*Andrena*) or leafcutter bees (*Megachile*) may fly 400 or 500 yards (365–457 m).
* Smaller bees, such as sweat bees (*Halictus*) and small carpenter bees (*Ceratina*), generally fly no more than 200 yards (182 m) from their nest.
* As for *Perdita*, the tiniest bees of all, they may be limited to only a couple hundred feet.

The shorter the distance a bee has to fly to find flowers, the more efficiently it can forage and provide for more offspring.

Some bees will also forage for nest-building materials. For example, mason bees search for areas of damp clay soil from which they gather and carry balls of mud back to their tunnel nests. Other bees gather plant resins or pebbles for the same purpose. Leafcutter bees search for leaves (and occasionally flower petals), often of very specific plants, from which to cut pieces to wallpaper their brood cells (see page 34). For example, some leafcutters seem to prefer cherry or snowberry leaves, or rose leaves or petals. One species known to visit cranberries on the East Coast of the United States preferentially cuts pieces from red maple leaves and typically is found only where red maple trees occur.

Wood- and Tunnel-Nesting Solitary Bees

Approximately 30 percent (around 1,200 species) of solitary bees in North America are tunnel nesters. Many tunnel-nesting species use abandoned beetle burrows in standing dead trees or limbs, but some chew out the soft central pith of dead, dry stems and twigs from such plants as elderberry and blackberry. A few have specialized needs: the inside of particular abandoned insect galls, for example, or the hollow stems of the common reed (*Phragmites australis*). What all of these nesting sites have in common is that they are dry, relatively warm, and offer protection from predators and parasites.

Most tunnel-nesting species divide the tunnel into a series of brood cells with materials they collect, such as leaf pieces, leaf pulp, tree resin, or mud. Leafcutter bees (genus *Megachile*) use pieces of leaf or flower petal to create self-contained brood cells. Using their mandibles, they cut particular sizes and shapes to fit different parts of the brood cell, lining the entire cell.

△ *The seal of a nest tunnel shows what type of bee is occupying it: leafcutters use leaf pieces, mason bees (such as this* Osmia sp.*) use mud.*

Unlike leafcutter bees, most tunnel-nesting bees do not line the entire cell, but simply build dividing walls across the nesting tunnel to create separate brood cells. Mason bees (genus *Osmia*) make these walls with mud or leaf pulp. Large carpenter bees (genus *Xylocopa*) and small carpenter bees (genus *Ceratina*) use wood fibers scraped from the walls of the tunnel to form dividers of "particle board." Other species produce building materials from glands in their own bodies. Yellow-faced bees (genus *Hylaeus*), for example, divide the tunnel into cells using a cellophane-like substance secreted from special glands. Each of these bees seals the nest entrance when it is finished with the same materials they used to construct the inner partitions.

△ *Yellow-faced bees (genus* Hylaeus*) use a cellophane-like glandular secretion to divide their nests.*

Ground-Nesting Solitary Bees

Most of North America's native bee species (about 70 percent, or very roughly 2,800 species) are solitary ground nesters. Ground-nesting bees dig nests in bare or sparsely vegetated soil that, depending on the species, varies from flat ground to vertical banks. Nest configurations range from a single short tunnel to complex, branching tunnel systems.

Rather than collecting materials from outside the nest with which to line their brood cells, many ground-nesting bee species smooth the cell walls with their abdomens and then apply a waxy or oily substance — produced from special glands near their mouths or on their abdomens — to line the cells, thus stabilizing the soil and protecting their brood. The substance lining the cell usually soaks into the soil, making it look shiny and helping to keep excess water out and microbes in check. The polyester bee (genus *Colletes*) secretes a cellophane-like material from abdominal glands to create a complete waterproof lining for the cells. Some ground-nesting species, such as tiny *Perdita* bees living in the southwestern deserts of the United States, leave the cells unlined. When the cell is complete, the bee seals it before beginning another.

△ *Polyester bees (genus* Colletes*) dig a narrow tunnel in which they create a series of waterproof brood cells.*

SOCIAL BEES

Social bees live in **colonies**, defined as having at least two adult females — there may be many more — that live in the same nest and share the work of preparing and provisioning it. Usually, one is the egg-laying queen and the others are workers that help construct, provision, and defend the nest.

European honey bees (*Apis mellifera*) are the first social bees most people think of, with their caste system of a single queen, multiple female workers, and a few male drones whose only purpose is to mate. Bumble bees (genus *Bombus*) are the best-known North American natives. There are 47 species of bumble bees in North America, including six social parasites that take over the nest of the other bumble bees. (Until recently, the parasitic species were in their own genus, *Psithyrus*. *Psithyrus* is now considered a subgenus of the genus *Bombus*.) There are also 200 or more species of sweat bees in several genera (mostly in *Halictus* and *Lasioglossum*) that sometimes nest socially.

△ *Hexagonal honeycomb crawling with bees is the stereotypical image of a social bee.*

△ *Honey bees are efficient foragers, collecting nectar and pollen from a wide range of flowers.*

up to 3 miles (5 km) to collect pollen and nectar — and twice that far to reach a particularly good source of nectar.

HONEY BEES AND HUMAN BEINGS

Because honey bees store surplus honey and produce beeswax, humans who desire these products have enjoyed a close relationship with this insect since earliest recorded history. Ancient rock paintings depict honey hunters scaling cliff faces and tall trees to harvest comb. Eventually humans learned to domesticate honey bees, and to keep them in artificial hives. By the first century A.D., Roman scholars had written instructions on how to construct hives. At some points in history, beeswax was actually the most prized product from the hive; it could be used for a wide variety of purposes. In fifteenth-century Europe, high-ranking household staff of the aristocracy often were paid partially in candles.

With domestication came the ability to move honey bees around the world, including to North America where no honey bees existed prior to their introduction by European settlers in the early 1600s. The European honey bee quickly became feral. Through swarming and through movement by humans, it spread across the continent and thrived for several centuries.

HONEY BEE DECLINES

In the past few decades, however, the United States has seen a dramatic decline in honey bee populations, resulting from the accidental introduction of exotic mites and diseases associated with Asian honey bee species. These parasites and the diseases they spread resulted in an almost complete disappearance of feral honey bees during the late 1980s to mid-'90s. In 2006, a new phenomenon emerged called Colony Collapse Disorder, which is indicated by rapid population declines in affected colonies. These threats, combined with decades of stagnant honey prices, have contributed to the 50-percent decrease in the number of managed honey bee hives in the United States since the late 1940s. The notable exception to this decline has been the spread of the Africanized honey bee in the southwestern states.

HONEY BEES AND NATIVE BEES

Because honey bees are an introduced domestic species that did not evolve with the native flora and fauna of North America, their conservation is beyond the scope of this book. The same efforts that support native bees will benefit honey bees, although researchers suspect nonnative honey bees are partly responsible for the declining populations of native bee species.

Over the past several years, studies have sought to determine whether honey bees forage at the expense of native bees. Few, if any, areas of North America are without honey bees, so comparative studies with sites that are free of honey bees are nearly impossible to conduct. In addition, scientists studying insect ecology must contend with significant natural variations in climate, available pollen and nectar, and usable nest sites, as well as other constraints on populations of native bees such as disease, predation, and parasitism. All of these factors may affect the

> The same efforts that support native bees will benefit honey bees, although researchers suspect nonnative honey bees are partly responsible for the declining populations of native bee species.

△ *Parasitic mites, like the one on the back of this bee, are just one of the causes of honey bee declines.*

number of native bees in an area, making it difficult to isolate the extent to which honey bees may be responsible for native bee declines. However, recent research does suggest that honey bees can significantly reduce reproduction in bumble bee colonies.

Workers in just a single hive of honey bees may collect hundreds of pounds of nectar and tens of pounds of pollen in the course of a year. Although scientists cannot confirm that the supply of nectar or pollen is limiting the reproductive success of native bees and other pollinators, evidence indicates that an abundance of honey bees reduces native bee populations in the same area if there is limited forage available.

Finally, honey bees are less effective than many native bees on a bee-for-bee basis at pollinating some plants, including various high-value crops such as blueberry, cranberry, tomato,

A NOTE ABOUT STINGS

WHEN WORKING AROUND BEES (and wasps) there is always a risk of getting stung. Most bees and wasps will not sting unless they are provoked, which is very difficult to do when they are visiting flowers. Many bees do not even have stingers that can penetrate human skin, although a small number of sweat bees and other solitary bees may sting if they become pinched or trapped against your skin.

Social bees and wasps, such as bumble bees, honey bees, yellowjackets, and hornets, are often defensive at their nests, but not when they are out foraging. Solitary native bees don't even defend their nests because doing so risks the ability of females to lay enough eggs to pass their genes on to the next generation. When any bee defends her nest or hive, she likely will be killed in the process. This is not a problem in social species where the queen stays inside the nest and can continue to lay eggs. The risk of dying and, therefore, preventing egg-laying is enough for solitary female bees to have evolved to avoid threats, rather than defend their nests.

Should you be stung by a honey bee, the only species in North America that leaves its stinger behind, be sure to remove the stinger as quickly as possible. Should you be stung by a native pollinator, there will not be a stinger to remove, but calmly move away from the site as you most likely are close to a nest.

Most people have mild reactions to bee and wasp stings and exhibit a reaction only at the site of the sting. However, it is important to monitor yourself after a sting for signs of a more severe reaction. Symptoms of a serious reaction include swelling elsewhere on the body, vomiting, dizziness, hoarseness, thickened speech, or difficulty breathing. If you experience any of these, see a physician promptly for medical care.

and alfalfa. Thus, such plants may set fewer seeds when visited by honey bees than they would if visited by a diversity of native bees.

In light of the uncertainties about the impact of honey bees on native plant communities and native bees, caution is best. Managers of natural areas should discourage the placement of honey bee hives, especially where the habitat is unique or rare. This restriction in a few rare habitats won't greatly impact the overall beekeeping community, and it may help conserve rare species of native bees, as well as the rare and endangered plants that depend upon native pollinators.

Bee habitat should be cultivated and protected in agricultural settings, which for years have depended upon the timely placement of honey bee hives in order to produce high yields. Researcher Claire Kremen of the University of California has studied the important roles of native bees and their habitat in agriculture, and concluded that their contribution could become "a valuable insurance policy if honey bees become scarcer or fail altogether."

Bumble Bees (family Apidae)

Bumble bees (genus *Bombus*) live in annual colonies founded in the spring by an individual queen after she awakens from hibernation. New nests are typically located within a dry cavity such as an abandoned mouse nest, a cavity in a tree, or under a tussock of grass. However, these adaptable bees will also nest in the walls of buildings, in the feathers and nest materials inside used birdhouses, in the jumble of an old rock wall, or even inside discarded mattresses and junked car seats.

The queen produces wax from glands in her body to make a cluster of various-sized round pots, lays her eggs in these pots, and then forages for nectar and pollen to feed her first brood of larvae. Unlike honey bees, none of the nectar that bumble bees store in these pots is converted to honey, and the queen only collects a small amount — just enough to provide a few days' worth of food in case inclement weather prevents foraging.

▽ *There are nearly four dozen species of bumble bees in North America.*

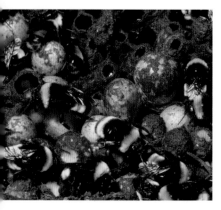

In addition to these wax pots, occasionally the entire nest will be covered with a canopy of wax called an **involucrum**.

Bumble bees can regulate their internal body temperature, and on cold days between foraging trips, the queen may sit atop the eggs, shivering to generate heat and incubating them much as a bird does. After a month or so, the larvae pupate and emerge as worker bees. These worker bumble bees then assume the work of expanding the nest and foraging, allowing the queen to stay safe inside and to use her energy for egg laying.

△ *Bumble bee nests contain a jumble of open pots, made from wax, for nectar storage and enclosed brood cells.*

LIFE CYCLE

The rapid development of bumble bees compared to solitary bees allows the colony to grow in a series of overlapping

LIFE CYCLE OF A TYPICAL BUMBLE BEE COLONY

1. A queen emerges from hibernation in spring and finds a nest site, such as an abandoned rodent burrow.

2. She creates wax pots to hold nectar and pollen, on which she lays and incubates her eggs.

4. In autumn the colony produces new queens and male bees, who leave to find mates. Newly mated queens hibernate and the rest of the bees die.

3. When her daughters emerge as adults, they take over foraging and other duties.

worker generations through the spring and summer. Late in the season, the colony switches from producing workers to producing new queens and males. At its peak, a colony may range in size from a couple of dozen to a couple of hundred bees. In extreme cases, some colonies have contained up to 800 bees.

REPRODUCTIVE LIFE

The queen can control the gender of her offspring: if she fertilizes an egg it becomes a female, if she doesn't it becomes a male. Most of her offspring are female workers. As mentioned, however, the colony will also produce a new generation of reproductive individuals, the males and a small number of new queens. Exactly what triggers the production of the males and new queens is not certain. A number of factors probably contribute, including the size of the colony, the ratio of workers to developing larvae, the abundance of the food supply, and seasonal cues such as day length. In some species, the queen releases pheromones to control the colony and mark larvae destined to become queens. New queens spend more time in larval development than do workers, growing larger and consuming more food.

After emergence, males quickly depart the nest to search for mates. They do not return. New queens remain in the nest as protection against predators; fed by foraging workers, they leave only to mate.

As fall arrives, most of the bumble bees — including the old queen — die. Only the newly mated queens survive the winter after leaving their natal nest, by digging several inches underground or beneath a layer of leaf litter to hibernate until the following spring, when they will establish new colonies.

Social Sweat Bees

A number of species in the sweat bee family (Halictidae) are social, though it can be difficult to separate them from their solitary bee cousins. Both social and solitary sweat bees nest in underground tunnels where females excavate cells and provision them with pollen and nectar to feed developing brood. To compound this difficulty, examples of social sweat bees can be found in a range of genera that are typically solitary ground-nesting species, including *Halictus, Lasioglossum, Augochlora,* and *Augochlorella*.

> ### HUMAN INTERVENTION AND SUPPORT
>
> Because only the final generation from each colony produces the queens for the next year, disrupting bumble bee nests early in the season before the colony has "reproduced" has especially dire consequences. Furthermore, some evidence suggests that less than half of bumble bee nests survive long enough to reproduce, making efforts to support bumble bee colonies by providing flowers throughout the growing season particularly meaningful.

△ *Sweat bee (genus* Lasioglossum*).*

△ *Male sweat bees are produced late in the life of the colony.*

The social organization of these sweat bee nests varies greatly depending upon the species, or even within a single species. In many cases, the nest cycle is similar to that of bumble bees (even if the nests themselves are not). A mated female bee emerges after winter hibernation and digs a burrow in which she excavates separate brood cells, provisions them, lays her eggs, and rears her offspring, who then become her workers. These workers help her continue to rear additional offspring over the summer months, ultimately leading to the production of males and the females that will hibernate and then start the next year's nests.

Sometimes, however, two or more sister females start a communal nest in the spring. In this case, one of these females lays eggs that the others tend. Usually one female is a dominant queen in the nest, rarely foraging, eating eggs laid by other females, and generally controlling their behavior. Often, the other females in such a nest are not mated or do not have developed ovaries and are smaller.

These examples of social behavior in sweat bees may be confused with a third type of relationship in which several females share a communal nest. In this case, each female does all of her own work to excavate and provision cells for her offspring, and the female bees are simply sharing the nest site and entrance.

Foraging of Social Bees

Social bees are generalists. To sustain the colony and feed the larvae through the spring and summer, they must gather food from flowering plants over an extended period. Individual bees typically visit a single species of flower on any given foraging trip, but they can switch to different flowers in response to seasonal and habitat changes.

KEY POLLINATORS

Bumble bees are often the first bees active in spring and the last bees active in the fall. Early-blooming plants such as willows and late-blooming plants such as goldenrod are especially important to their survival. Their ability to regulate body temperature by shivering or basking in the sun (easily raising their temperature up to 100°F) enables bumble bees to forage during wetter, cooler conditions than honey bees and many other native bees. One species, *Bombus polaris,* lives in northern Canada, Alaska, and Greenland where few other bees are found. During the short summers, *B. polaris* forages around the clock in the 24 hours of daylight. This capacity to work in cold weather makes bumble bees important pollinators of spring-blooming wildflowers and fruit crops.

Another feature that makes bumble bees important pollinators is their unusual ability to **buzz-pollinate** (also called **sonicate**) flowers by disengaging their wings from their flight muscles, and using those muscles to shake their entire body at a frequency close to a middle C musical note. This vibration significantly increases the release of pollen from some flowers, including tomatoes, peppers, blueberries, and cranberries. Several other bees also have the ability to perform buzz-pollination.

WAYS OF CARRYING POLLEN

DIFFERENT BEES have different adaptations for transporting pollen. Bumble bees, like honey bees, have a concave depression on the upper part of their hind legs surrounded by stiff hairs. This structure, called a **corbicula** or **pollen basket**, serves to hold packed clumps of pollen grains gathered during foraging. The pollen is moistened with nectar to help hold it in place.

Sweat bees, in contrast, carry dry pollen, packed into patches of stiff hairs on their hind legs. These hair patches are called **scopae** or **pollen brushes**. Sweat bees are much smaller than bumble bees and thus carry much less pollen, but their abundance makes them important pollinators in any ecosystem.

Bumble bees will forage up to a mile from their nests, their large size allowing them to exploit widely spaced flower patches. Social sweat bees are far smaller and can't fly nearly as far. Obviously, since sweat bees vary in size, their foraging range does too. In general, they are limited to flying a few hundred yards from their nests.

Parasitic Bees

Bees are generally perceived as hard-working creatures; "busy as a bee" is an oft-heard saying. Not all bees, however, fit this perception. About a quarter of all bee species are parasites: they lay their eggs in the nests of other bees and allow the other species to do all the hard work. Female parasitic bees do not collect their own pollen, so they lack any special pollen-holding structures. They are occasionally hairless and wasplike in appearance.

SOLITARY AND SOCIAL PARASITES

There are two types of parasitic bees. The majority of parasitic bees are **cleptoparasites**: solitary bees that lay their eggs in the nests of other solitary bees. They are commonly called cuckoo bees, after the well-known birds that lay their eggs in other's nests. The solitary cleptoparasite enters the host's nest to lay her eggs when the female of the host species is not present. Most species simply lay an egg and leave.

The first act of the cuckoo bee larva after hatching is usually to kill the host bee's egg or larva. (One of the few exceptions to this is the genus *Sphecodes*, whose females will destroy the host egg before they lay their own.) After hatching, the parasitic larva then feeds on the host's food provisions, and it will remain in the brood cell through larval and pupal stages before emerging as an adult.

The second type is the **social parasite**. A social parasite enters the nest of social or semisocial species, replaces the queen, and has the workers rear her offspring. Cuckoo bumble bees are the best example of social parasites in the United States and Canada. These bees (genus *Bombus*, subgenus *Psithyrus*) do not have the ability to construct their own nests or to collect pollen. Instead they enter established colonies of bumble bees where they either kill or subjugate the queen. The cuckoo bumble bee then controls the existing workers (including the

---

PARASITES CHOOSE THEIR HOSTS WELL

Not surprisingly, parasitic bees require specific hosts. Bumble bee parasites will use only a few species of bumble bees as hosts. For example, Ashton's cuckoo bumble bee (*Bombus ashtoni*) parasitizes the rusty-patch and yellow-banded bumble bees (*B. affinis* and *B. terricola*, respectively), two closely related species from the eastern United States. Cleptoparasites similarly exploit related species in the same family as themselves; *Coelioxys* in the leafcutter family (Megachilidae) only attacks other leafcutter bee nests.

queen if she is still alive) using pheromones or physical attacks, causing them to raise the parasite's larvae as if they were their own.

Upon hatching, the new cuckoo bees disperse, mate, and then enter hibernation. Most cuckoo bumble bee adults emerge from hibernation later than their host species. This delayed timing allows the host to establish her nest and raise a first generation of workers. Although this type of nest parasitism may seem detrimental to bumble bee populations, the presence of cuckoo bees usually indicates healthy host populations, and consequently healthy ecosystems.

△ *Cuckoo bees such as this* Triepeolus *are cleptoparasites in the nests of solitary bees.*

WASPS

Wasps are the closest relatives of bees. From an evolutionary standpoint, bees are simply wasps that have adopted a vegetarian diet. Wasps in contrast are typically carnivores during the larval stage, feeding on insect prey provided by their mothers or bits of meat scavenged by their mother from animal carcasses (and people's picnics!). During the adult stage, when a protein-rich diet is no longer required to grow their bodies, wasps switch to a carbohydrate-based diet and are typically fueled by sugary flower nectar, rotting fruit, or even soda.

Wasps as Foragers

Unlike bees, wasps typically have short tongues and are unable to reach the deep hidden nectar reservoirs of some flowers. Consequently, nectar-collecting wasps are most frequently observed on shallow flowers with readily accessible nectar, such as goldenrod, or members of the Carrot family, such as fennel, Queen Anne's lace, and parsnip. Wasps also frequently seek out nonfloral sugar sources such as **honeydew** (the sugary excrement of aphids and scale insects), rotting fruit, or the extrafloral nectaries found at the base of leaf petioles on plants such as leafy spurge or partridge pea.

△ *With a few notable exceptions (such as bald-faced hornets and yellowjackets), wasps are gentle insects that do not typically sting people. Wasps make important ecological contributions by controlling pest insects.*

Many wasps are smooth-bodied and do not actively collect pollen. Those with hairs lack the branched, pollen-trapping hairs found on most bees, making them relatively minor pollinators of most plants. Nonetheless, they do provide some incidental pollination, carrying and dropping a few pollen grains as they move among flowers.

The following descriptions include some of the wasps most frequently encountered on flowers.

Yellowjackets, Hornets, and Paper Wasps (family Vespidae)

Yellowjackets (genus *Vespula*), hornets (*Dolichovespula*), and paper wasps (*Polistes*) comprise approximately 40 native and exotic species in North America, collectively referred to as **social vespid** wasps. Most of these wasps are true social insects, with multiple females cooperating to construct and defend a nest and to rear the offspring. Often these nests consist of conspicuous paper disks constructed beneath the eaves of houses, or enclosed paper structures hanging in trees.

When threatened, social vespids will defend these nests aggressively with a succession of painful stings. This habit, along with their common appearance at picnics to scavenge food, including sugary soft-drinks, makes them familiar to anyone who spends much time outdoors.

Both yellowjackets and hornets can be recognized by relatively stout bodies with distinct yellow-and-black or white-and-black coloration. Paper wasps, in contrast, have more elongated bodies (up to 1 inch [25 mm] long), slender waists, and less-distinct color contrasts.

△ *Yellowjackets are commonly encountered at summer outdoor gatherings.*

▽ *Although sometimes aggressive when defending their nests, paper wasps are surprisingly gentle while drinking nectar.*

NONNATIVE WASPS

SOME ECOLOGISTS have expressed concern about the effects of nonnative social vespids on native North American ecosystems. For example, some speculate that nonnative *Polistes* wasps prey extensively on butterfly caterpillars, including rare species, to feed their offspring. In addition, the introduced wasps may occupy cavities normally used by other wildlife, including hollow tree cavities and bluebird houses. These issues should not be interpreted as an excuse to reach for a can of insecticide, but they are a reminder of the potential impact that exotic species can have on native ecosystems. These wasps were not intentionally introduced to North America; they likely arrived by accident on ships from Europe.

Despite these concerns, native vespid wasps, and potentially some of the nonnative ones, prey on a number of crop pests, and their presence can help reduce the need to use insecticides in vegetable gardens and on farms. Because it can be difficult to distinguish native social vespids from nonnative ones, the best course of action is simply to leave them alone.

Potter Wasps (family Vespidae)

Potter wasps comprise approximately 250 species within the Vespidae subfamily Eumeninae. These solitary wasps are primarily caterpillar hunters that construct tubular mud nests on the sides of protected stone surfaces or brick walls. They may also occupy existing wood or soil cavities identical to those used by solitary bees.

Eumenine wasps vary in size, shape, and color. Common species range from 0.5 to 1 inch (10–25 mm) long, have slender waists, and sport white, yellow, or even orange markings. Common North American genera include *Ancistrocerus, Eumenes, Euodynerus, Odynerus, Pterocheilus, Stenodynerus,* and *Symmorphus.* Like the social vespids, potter wasps can benefit humans by preying upon such pests of forests and farms as the spruce budworm and alfalfa weevil.

Pollen Wasps (family Vespidae)

The nonsocial Vespidae subfamily Masarinae includes about a dozen North American species. These are the only wasps that visit flowers to collect nectar and pollen as food for their young. In contrast to other vespids, pollen wasps have longer mouthparts than most wasps, which help them reach nectar from deeper-throated flowers. The genus *Pseudomasaris* includes all the North American pollen wasps, most of which are restricted to the western United States. *Pseudomasaris* generally resemble small yellowjackets (0.4 to 0.8 inches [10–20 mm] long) with yellow and black stripes and a distinctive knob at the end of each antenna. Although *Pseudomasaris* may visit a wide variety of flowers for nectar, they seem to prefer pollen from only two plant families: the Waterleaf and Figwort families.

Digger Wasps and Sand Wasps (family Sphecidae)

The digger and sand wasps include about 1,200 species in North America and are the closest wasp relatives of bees. Sphecid wasps are an extremely diverse group of solitary hunters with great variation in body size and form. Examples include the tiny and slim aphid-hunters (genus *Passaloecus*) that often nest in narrow holes in wood, and the large cicadakillers (genus *Sphecius*) that may grow to 1½ inches (40 mm) long. Sphecid wasps also vary in color, from solid black to combinations of white, yellow, orange, red, metallic blue, and green.

△ *Sphecid wasps specialize in catching other insects to supply their nest.*

Spider Wasps (family Pompilidae)

Spider wasps are distinguished by long slender legs and an unusual habit of constantly twitching while walking. Typically solid black in color, spider wasps often have a metallic blue sheen or occasionally sport white, yellow, or red patches on their bodies, and may have orange or gray wings. Most spider wasps range from 0.4 to 1 inch (10–25 mm) long. There are approximately 220 species in North America.

Some of the most notable spider wasps are the exceptionally large (up to 1½ inches, or 40 mm) tarantula hawks (genera *Pepsis* and *Hemipepsis*) of the southwestern states. These powerful hunters, while not aggressive toward humans, reportedly deliver some of the most painful stings of any North American insect when handled. When foraging for nectar, tarantula hawks are found on flowers including milkweeds and creosote bush.

Other Wasp Families

In addition to the wasp families already summarized, North America is home to a rich diversity of other, less common wasps often found visiting flowers for nectar. Among these are the approximately 435 species of colorful orange, pink, red, blue, and yellow insects known as velvet ants, which are actually wasps in the family Mutillidae. Of the 200-plus species in the Tiphiidae family, those in the genus *Myzinum* are the largest and most colorful. Adults are regularly found on flowers. Like all tiphiids, their larvae are parasites of beetle larvae.

Another wasp family whose larvae parasitize beetle larvae is the Scoliidae. Scoliid adults are large — up to 2 inches (50 mm) — and hairy. Males of some species sleep in clusters, with their bodies tightly curled around a plant stem. The Chrysididae family is a somewhat large group with roughly 230 species. It includes brightly colored metallic green or blue species that lay their eggs in the nests of other wasps and bees, consuming the host's food provision and often the host's larva. This nest parasitism has earned them the common name of cuckoo wasps.

△ *Chrysidid wasps are often called cuckoo wasps because they lay their eggs in the nests of solitary wasps or bees.*

FLIES

Flies belong to one of the largest insect orders on earth, the Diptera, which includes 188 families and nearly 120,000 species worldwide. Among their ranks are tiny species barely visible to the unaided human eye, and species more than an inch long. Flies are among the most frequent visitors to flowers; 71 different families have been recorded visiting flowers. Some of these species are pollinators; some are not.

With their reputation as generalist foragers, no nests to provision, and sometimes sparsely haired bodies, flies don't get much credit as significant pollinators. Despite this reputation, they are often important pollinators in natural ecosystems for specific plants, and occasionally for human food plants.

Flies also are often misidentified as wasps or bees, and for good reason. Flies can't sting, but by sometimes mimicking stinging insects they make birds and other predators avoid them. Syrphid flies, for example, gain protection by mimicking the bold warning (**aposematic**) coloration of bees and wasps.

LIFE CYCLE AND FEEDING

Depending on the species, an adult fly's life span varies from a few hours to several weeks. Adult flies don't make nests but instead lay their eggs close to (or within) a good supply of the larvae's preferred food, which varies greatly among fly species. The larval food of syrphid species, for example, includes detritus and aphids; the larvae of bee flies are parasites in bee nests; and small-headed fly larvae are internal parasites of spiders.

POTENTIAL AS POLLINATORS

The search for a suitable place to lay eggs requires energy, which many flies derive from flower nectar; in the process of feeding they act as pollinators, albeit generally poor ones. The groups of species we describe here are the best pollinators among the fly species.

Flower Flies (family Syrphidae)

Flower flies — or hover flies — are among the most colorful and conspicuous flies found around flowers. Some species regularly hover above flower heads, and many eat nectar or pollen as adults. Many flower flies mimic the coloring of bees or wasps, an adaptation that enables them to confuse predators. For some species, this mimicry confers more than just

FLY OR BEE?

△ *Syrphid fly (genus* Eristalis)

△ *Leafcutter bee (genus* Megachile)

TO DISTINGUISH a fly from a bee, first examine its head. Most flower-visiting flies have short, fat, down-turned antennae (often looking like misplaced earlobes with obvious bristles emerging from the top); bees' antennae tend to be much longer and more uniform in thickness. Second, look at the wings. Flies have two — the name Diptera means "two wings" — and bees have four. Finally, look for body features that enable the insect to carry pollen on its legs or its belly; female bees generally have such features and flies do not. Some of these characteristics may be hard to discern on an insect feeding at a flower, but with practice you can learn to see them at a glance. (See also page 227.)

△▷ *Different types of flower flies can look very similar, for example, the genera* Syrphus *(left) and* Eupeodes *(right). Both are widespread and can be found throughout North America.*

protection while they are foraging. Flies in the genus *Volucella*, for example, make use of mimicry to enter the nests of bumble bees, where they lay their eggs. They are not parasites on the bumble bees; their larvae act as nest cleaners, feeding on dead bees and other detritus.

The larvae of most flower flies are predators of other insects. Look closely at a rose bush, and you may see a grublike flower fly larva rearing up to consume an aphid. (They typically feed at night, so you may have to go out after dusk with a flashlight!) Their diet of other insects makes flower flies valuable for pest control as well as pollination. Other species live in ant nests where they feed on ant brood, or lay eggs on rotting wood, into which the larvae tunnel as they eat wood fiber. Larvae of the widespread drone fly (*Eristalis tenax*), known as rattailed maggots because of the long breathing tube at their rear ends, can be found in pools of bacteria-rich liquids, such as manure piles and stagnant puddles.

△ *Syrphid fly larva eating an aphid.*

Bee Flies (family Bombyliidae)

The bee flies are another important group of pollinators. Many of the nearly 300 species are stout and hairy, like bumble bees, or are striped with black, white, or shades of brown. They also drink nectar from flowers, reinforcing the resemblance to bees that gives them their name. A number of species have a prominent bristlelike proboscis to reach the nectar deep within flowers, and others can hover while feeding. Most species have banded or spotted wings that stick out sideways when the insect is at rest.

nectar to fuel this effort and to keep themselves warm. Flies, in contrast, don't have a nest to supply, so require less energy. Flies also can spend more time basking on flowers, their dark bodies absorbing the sun's rays for heat. As a result, flies require much less nectar to survive and reproduce in cold climates and can be more common pollinators than bees in these areas.

BUTTERFLIES AND MOTHS

Butterflies, possibly the best loved of all insects, are appreciated as benign creatures that add color, beauty, and grace to our landscapes. Butterflies and moths belong to the same insect order, Lepidoptera, and while people can usually distinguish a butterfly from a moth, the two can sometimes be hard to tell apart (see box).

BUTTERFLY OR MOTH?

△ *Tiger moth* (Gnophaela vermiculata)

△ *Variegated fritillary* (Euptoieta claudia)

IN GENERAL, butterflies are brightly colored and fly by day, and moths are more likely to be colored in muted grays and browns and fly at night. But there are numerous exceptions. Such moths as the burnets, foresters, tigers, and ctenuchas are day-flying and colorful, as are some of the hawk moths, while skippers, on the other hand, are small, brown mothlike butterflies.

When moths and butterflies are at rest, it is easier to see other distinguishing morphological and behavioral differences. Butterflies tend to hold their wings either partially open or closed vertically over their bodies, like the sails of tiny sailboats. Most moths, on the other hand, hold their wings flat, with the forewings covering the hind wings. A closer look reveals that moths tend to be fatter and hairier than butterflies. Antennae also differ: a butterfly antenna is a single filament with a clubbed tip (which on many skippers is bent or hooked), whereas a moth antenna may be broad and feathery (usually males), or a single filament that tapers to a point (females).

The Lepidoptera is a diverse order, containing more than 12,000 species in North America. Moths make up most of this diversity; there are only 800 butterfly species north of Mexico.

LIFE CYCLE

Both butterflies and moths begin life as eggs laid on or near the species' particular host plant. Each hatches as a tiny, soft-bodied caterpillar, eating and growing until it transforms into a pupa or **chrysalis**, the mummy-like quiescent stage between larva and adult. "Chrysalis" is the name of both a stage in development and the encasing at this stage. Many moths also spin an additional layer of protection: from their silk glands they make a **cocoon**, which surrounds the chrysalis. During the pupal stage the structure of the caterpillar is reorganized as it completes the metamorphosis into a winged adult.

△ *Butterflies pupate from caterpillar to adult as a chrysalis.*

Butterflies

Although butterflies are not the most important pollinators of plants, they are among the most conspicuous. They float like kites across an open space or garden, with their wings flashing bright yellow and black, brilliant orange, or in subdued shades of blue or copper. For centuries, butterflies have attracted the attention and admiration of people across the world. Indeed pollinator gardening started as an attempt to lure them to yards, where they delight young and old alike. Many thousands of people own books dedicated to identifying butterflies and telling their stories. Venturing far beyond their own backyards, some spend their vacations at eco-resorts geared toward offering tourists glimpses of large, spectacularly colored equatorial species.

△ *Fender's blue (*Plebejus icarioides fenderii*)*

FEEDING

Like most wasps and flies, butterflies consume nectar only as adults, to fuel their own flight to find mates and lay eggs. They do not feed on or actively gather pollen, nor do they provision nests for their young. Instead, most butterflies lay their eggs on just a few closely related plants. After hatching, the caterpillars eat only the leaves or flowers of those plants. Consequently, butterflies lack the branched pollen-trapping hairs found on bees. And while butterflies are sometimes fuzzy enough for pollen to stick to them, they frequently also have very long tongues that allow them to probe deeply into flowers

△ *Common buckeye (*Junonia coenia*)*

for nectar, often bypassing the anthers. Nonetheless, through the occasional accidental dusting of pollen, they do contribute to plant reproduction.

HABITAT NEEDS

Like most insects, butterflies need warm temperatures and sunshine in order to fly. Consequently, they are often found flying in open areas or those with good southern exposure. In the early morning, butterflies bask on sun-warmed rocks, bricks, or gravel paths. As the morning warms, they begin visiting their favorite flowers for nectar, preferably in sunlit areas. They also prefer areas that are sheltered from prevailing winds, as most are weaker flyers than many other insects. Some butterflies, such as monarchs, may fly long distances by soaring on wind currents; others, like skippers, are not strong flyers.

Along with the need for sunny locations and specific host plants, many butterflies have other subtle habitat requirements. For example, many species of the gossamer wings (the family Lycaenidae) are **myrmecophilous** (ant loving).

PROTECTING BUTTERFLIES

LIKE MANY INSECTS, butterflies are threatened by urbanization, the intensification of agriculture, the widespread use of pesticides, and the introduction of invasive species that out-compete native host plants. However, they are better studied than other invertebrate groups. NatureServe, a nonprofit network of natural heritage programs that tracks rare and endangered species, estimates that 17 percent of U.S. butterflies are vulnerable, imperiled, or critically imperiled, and the World Conservation Union estimates that 10 percent of swallowtails are considered threatened worldwide. Currently, there are 22 butterfly species formally listed as threatened or endangered under the U.S. Endangered Species Act, but that number is likely grow without significant efforts to halt their decline.

The good news is that you can take specific steps to protect what E. O. Wilson has called "the flowers of the air." All the habitat conservation efforts outlined in this book, including those more specifically tailored to bees, will benefit butterflies. The nectar plants featured in Part 4 include many species that will attract and support butterflies. Part 4 also contains a list of host plants for the most common butterflies in the United States and Canada, all of which are described at the family level in the following section. Many other books are already widely available to help with identifying individual butterfly species (see appendix).

◁ Skippers are often rather drab. Their antennae have a distinctive hooked tip, and many species at rest hold their wings in a characteristic double V position.

The caterpillars secrete honeydew, a sugary and sometimes protein-rich substance that attracts certain species of ants. These ants then tend the caterpillars and protect them from predators in return for the honeydew.

Skippers (family Hesperiidae)

Skippers are small butterflies, usually brown or orange-brown, with a characteristic rapid, erratic flight pattern. When stationary, they often hold their wings partially open with the forewing separated from the hind wing, thus making two Vs, one inside the other, when observed from the front or back. They also can be distinguished from other butterflies and moths by the hooked bulb at the end of each antenna. On many skippers, this bulb is bent almost 90 degrees to the side, rather than positioned in a straight line as it is on other butterflies. Moths lack this bulb entirely.

Swallowtails (family Papilionidae)

Swallowtails and their relatives include the largest butterflies in the world. Most species have tails on their hind wings (although those in the subfamily Parnassiinae do not), and many are black and yellow or black and white. The adults are strong fliers, sometimes patrolling back and forth along wooded edges or stream corridors. They also may be seen gathering together at mud puddles. As a means of defense, the young caterpillars of certain swallowtails mimic bird droppings with extraordinary accuracy.

△ Western tiger swallowtail (Papilio rutulus)

△ *Cloudless sulphur (Phoebis sennae)*

△ *Bronze copper (Lycaena hyllus)*

▽ *Monarch (Danaus plexippus)*

Whites and Sulphurs (family Pieridae)

As the name implies, these butterflies are usually white or yellow. They vary in size from tiny to large and are widely distributed throughout the United States. They fly in a continuous fluttering pattern and are conspicuous along roadsides, in meadows, and in gardens. In many cases, the white scales reflect patterns of ultraviolet light, invisible to us, but obvious to other pierids in search of a mate. Larvae of most whites feed on Mustard-family plants such as cabbage; those of most sulphurs feed on plants in the Pea family (legumes).

Gossamer Wings (family Lycaenidae)

This large family is relatively easy to identify since it includes most of North America's smallest butterflies: coppers, hairstreaks, blues, and metalmarks. Although small, the surfaces of gossamer wings are often brightly colored in blues, greens, or coppers and may be dotted with bold spots or checkers. Many of these butterflies also have hairlike tails and bright eyelike spots at the rear of their hind wings, intended to mimic the head and antennae. This adaptation is intended to fool predaceous birds into striking the hind wing area, leaving the rest of the butterfly to escape, torn but essentially unharmed. This is also the family of butterflies that as caterpillars attracts certain species of ants with honeydew secreted from a special gland.

Brush-Footed Butterflies (family Nymphalidae)

This family contains many of the best-known butterflies, such as monarchs and painted ladies. They are called brush-footed because the butterflies in this family have greatly reduced forelegs, covered in hair. These forelegs are so much smaller they appear to have only four legs. Because of their great diversity, no single field characteristic makes the members of this family easy to identify as a group. The nymphalids are often colored in shades of orange, brown, and black, sometimes checkered or dotted with eyespots or silver spots. Some are even cryptically shaped and patterned like dead leaves. The satyrs and the milkweed butterflies are two subfamilies of the Nymphalidae that are sometimes viewed as distinct families.

SATYRS (SUBFAMILY SATYRINAE)

Until recently, satyrs were classified as a distinct family. Members of this subfamily of brush-footed butterflies are medium-sized and often drab. They are usually various shades of brown and often have darker eyespots on their wings. They have a weak, bobbing flight and are often seen near woods, prairies, or alpine areas among grasses, which are their larval host plants.

MILKWEED BUTTERFLIES (SUBFAMILY DANAINAE)

This subfamily is known in North America by its most familiar member, the monarch, but also includes numerous other species, often large and all tropical. Monarchs use milkweeds as their sole host plants and are the only butterflies in the world that regularly migrate thousands of miles to overwinter. Poisons derived from their milkweed hosts make monarchs unpalatable for birds, a condition advertised by their bright orange and black design.

Moths

Of the insects described in this book, moths may be the group least likely to be noticed in the landscape, or to be the focus of many people's conservation efforts. The muted colors of many moths, their largely nocturnal lives, and the reputation of only a few species as crop or wardrobe pests results in their typically being overlooked at best or despised at worst.

This is unfortunate because there are more than 10,000 moth species in North America (compared to about 800 butterflies). Many of these are very important specialist pollinators and/or food for other wildlife such as songbirds, and they lead fascinating lives in their own right.

LIFE CYCLE AND FEEDING

The general life cycle of moths and butterflies is largely the same. As caterpillars, moths feed on a slightly larger range of food sources than butterflies do. Many require specific host plants, and a few eat seeds or other organic matter. Less than 1 percent of moth species (limited to the Tineidae family) eat fabric made from natural protein-rich fibers such as wool or silk.

△ *Tiger moth (Gnophaela sp.)*

commemorate this prediction, the moth was named *Xanthopan morganii* ssp. *praedicta*; the subspecific name means "the predicted one."

Closer to home, many North American sphinx and hawk moths are usually crepuscular, so the best time to observe them feeding is at dusk or dawn. In garden settings they are sometimes attracted to morning glories, various penstemons, and beebalm. You may see them feeding at hummingbird feeders. Viburnum, snowberry, and blueberry shrubs are some of the larval host plants for some sphinx and hawk moths.

Owlet Moths (family Noctuidae)

Owlet moths are a major group of flower visitors, especially the cutworm and the plusiine looper moths. Cutworm moths are typically small, drab-colored, and heavy-bodied moths that hover while feeding. Plusiines are similar in build and behavior but are distinguished by silvery marks on their forewings. The forewings of one plusiine, the green-patched looper (*Diachrysia balluca*), are a striking brassy metallic green.

Underwing Moths (family Erebidae)

Underwings are large moths whose dull-colored but intricately patterned forewings provide excellent camouflage. In contrast, their black hind wings have striking color bands of red, orange, or yellow that can startle when suddenly exposed. The larval food of underwings usually consists of tree leaves from willows, aspens, oaks, hickories, hawthorns, and others. A succession of species may be found in many areas during the summer, although their nocturnal nature makes them difficult to see. One way to observe underwings is to check at night with a flashlight for their presence at butterfly feeders baited with overripe bananas (see page 119, Feeding Butterflies). Like butterflies, underwing moths are attracted to the aroma of fermenting fruit.

△ *Moths in the genus* Drasteria *are known as lesser underwings.*

Geometer Moths (family Geometridae)

Geometer (or geometrid) moths are small, dainty, drab-colored fliers whose shape is similar to butterflies'. They are common flower visitors at night, when no butterfly would be out foraging. The name Geometridae derives from Latin and Greek words meaning "earth-measurer," a reference to the unusual locomotion of their caterpillars, the familiar inchworms.

Butterfly and Moth Feeding Habits

One of the remarkable things about butterflies and moths is that they begin life with biting mouthparts for chewing on leaves and they end it with mouthparts adapted only for sucking liquids. Thus, to complete their life cycles, they need very different food sources. As caterpillars they eat plants — known as **host plants** — and they can be very particular about which ones. Caterpillars of some species will eat only a single species of plant or a small number of closely related plants, whereas other species will eat a very wide range of plants from numerous families.

The endangered Karner blue butterfly (*Plebejus melissa samuelis*) feeds only on the wild lupine (*Lupinus perennis*), a member of the Pea family. The Karner blue, like other species that are highly selective in their food requirements, will not survive without the plant. Many species that rely on a single particular food source are threatened when the plant population is threatened. In the case of the Karner blue, the Great Lakes dune habitat supporting wild lupine — formerly found from the upper Midwest east to New York and New Hampshire — is now in drastic decline, thus jeopardizing the survival of the butterfly.

At the opposite end of the spectrum is the gray hairstreak, a butterfly found from coast to coast. Its caterpillars have been found eating more than 80 different plant species from at least 24 plant families.

Surviving Winter

Different species of butterflies and moths use different strategies to survive the winter or other extended periods of inclement weather. Most species pass the winter as eggs, caterpillars, or pupae in protected surroundings such as leaf litter or dense vegetation. A few, such as mourning cloaks and tortoiseshells (*Nymphalis*) and anglewings (*Polygonia*), overwinter as hardy adults. In natural conditions, these butterflies find shelter from predators and the cold in caves or tree cavities, under leaf litter, a pile of rocks, or among evergreen vegetation, but they will also seek winter refuge in barns, or even in a cool room in a house.

△ *Karner blue butterfly* (Plebejus melissa samuelis)

△ *Painted lady* (Vanessa cardui)

△ *Mead's wood nymph* (Cercyonis meadii)

bees, because the female bees construct their nests within abandoned beetle tunnels. A lack of sufficient tunneling beetle larvae may limit local bee populations.

Within the immense diversity of beetles, not all visit flowers, but there are several families of beetles that include regular flower visitors likely to be encountered by people in Canada and the United States. The most common are described here.

Soldier Beetles (family Cantharidae)

Adult soldier beetles have long, almost rectangular bodies that are often bright red and black (resembling military uniforms from precamouflage days) or yellow and black. The

elytra (hardened wings that form the protective casing over their backs) of soldier beetles are soft and flexible, giving rise to an alternative name, leatherwings. Adults are commonly observed on showy flowers, including sunflowers, coneflowers, and goldenrod, where they feed on pollen and mate. The larvae are highly mobile predators that hunt through leaf litter or under stones and debris for eggs or larvae of other invertebrates.

△ *A pair of Pennsylvania leatherwings* (Chauliognathus pensylvanicus) *on goldenrod*

Long-Horned Beetles (family Cerambycidae)

Long-horned beetles are also large, but they have antennae that may equal or exceed their body length. Adults of many species lay eggs on or within dying or weakened trees, where their grublike larvae burrow beneath the bark or through the wood. Because decaying wood is not a good source of nutrition, these larvae may take several years to develop fully, often relying on symbiotic microbes in their guts to digest cellulose in the wood.

A few long-horned beetle species consume live plant tissue during the larval stage, and a few are very specific in their host plant requirements. Milkweed beetles (genus *Tetraopes*), for example, feed only on the roots of milkweed plants as larvae.

As adults, many long-horned beetles are attracted to flowers and contribute to pollination. The locust borer (*Megacyllene robiniae*) is one common example, particularly on goldenrod blossoms in early fall. This conspicuous beetle is notable for its bee- or wasplike yellow-and-black color pattern.

Jewel Beetles (family Buprestidae)

These aptly named beetles include some of the most beautifully colored insects in North America; most adults feature flashy metallic coloration. Their bodies appear almost bullet-shaped, with a rounded head and thorax that seem to merge into a tapering abdomen. Their larvae are wood-borers, chewing out tunnels as they feed inside dead trees.

△ *Spotted tylosis (Tylosis maculatus)*

Blister Beetles (family Meloidae)

Blister beetles are usually black in color and possess long legs and an elongated thorax. Blister beetles can also be fairly large; a very common eastern North American species, the black blister beetle (*Epicauta pennsylvanica*), routinely reaches more than an inch (25 mm) long at maturity.

Blister beetles have a fascinating life cycle that begins as an egg laid on a flower. The newly hatched larvae (called **triungulins**) are highly mobile and attach themselves to a foraging bee. The larvae ride on the bee until it returns to its nest, where they drop into a brood cell, eat the egg, and then consume the pollen and nectar. Because of this unique life cycle, blister beetles are highly dependent on healthy native bee populations, and their presence can serve as an indicator that host bee populations are robust.

▽ *Blister beetles produce cantharidin when threatened, a chemical that may blister human skin.*

As adults, the beetles feed on nectar and pollen from flowers. In the case of some blister beetles, such as those in the genus *Nemognatha*, mouthparts are adapted to form a tonguelike structure able to reach within deep flower structures.

means that they will visit picnics for ripe fruit or fermented drinks, hence another common name, beer beetles. Larvae develop in decaying or composting materials of all types.

BEETLE FEEDING HABITS

A substantial percentage of beetles that feed at flowers eat pollen. Some even chew on the flower itself; this form of beetle interaction is called mess-and-soil pollination. Despite the damage they can cause with their chewing and rummaging, a little pollen sticks to the beetles' bodies and gets transferred from anther to stigma. Because of the abundance of these insects, even that small amount of pollen adds up, making beetles significant pollinators. In desert areas and humid tropics, the importance of beetles as pollinators is compounded by their sheer numbers and the emergence of adults in synchrony with the seasonal blooming of flowers, either in the spring or during the summer rainy period.

Threats to Pollinators 4

HUMAN BEINGS have been shaping and modifying the earth since the dawn of humankind. Some Native Americans, for instance, burned grasslands year after year to keep the forests from encroaching and thus created favorable habitat for the game and plants they harvested. Today, humans possess power to rapidly, profoundly, and permanently alter landscapes for agriculture, forestry, mining, and urban development. This habitat alteration can dramatically affect the survival rate of bees and other pollinators.

✿ Habitat loss reduces pollinators' foraging and nesting options.

✿ Introducing exotic organisms, accidentally or intentionally, can cause catastrophic ecosystem disruptions.

✿ Climate change may cause the extinction of some species and others to shift their ranges.

✿ Pesticides can kill pollinators directly, reduce reproduction, and kill the plants they need to survive.

△ *Native garden plants are usually adapted to local pests and typically need no insecticides to stay blooming and beautiful.*

△ *Pesticides directly harm foraging bees and can be carried back to the nest in toxin-tainted nectar and pollen.*

In 1973, in Washington state, the diazinon applied to control aphids on alfalfa also decimated foraging alkali bees, which are an important pollinator of alfalfa. The die-off of female bees led to a 95 percent drop in the number of underground bee larvae in three nearby nesting sites studied by one investigator.

These incidents are just three well-documented poisonings out of the many thousands that have occurred. The documentation of the New Brunswick bee kill is unusual in that it relates to wild bees. Far more often, the impacts of pesticides on managed pollinators are documented while the extent of the killing of wild bees goes unnoted.

Virtually all of the pollinator research on the effect of pesticides focuses on honey bees because of their importance to agriculture. However, the use of managed populations of solitary bees — in particular, alfalfa leafcutter bees (*Megachile rotundata*) and alkali bees (*Nomia melanderi*) for alfalfa — has provided opportunities and incentive to extend studies to native bees.

Urban and Suburban Pesticides

Pesticides are not just a problem on agricultural lands. Studies conducted by the U.S. Geological Survey and some municipalities have detected higher concentrations of pesticides in urban streams than in streams in agricultural areas. A study in the Puget Sound Basin of Washington found that more pounds of pesticides were applied per acre in urban neighborhoods than on farmland.

This should not be surprising. On farms, pesticides can be sprayed only by trained and licensed applicators, and the growers want to save money by using as little pesticide as they can. In addition, there are laws regulating pesticide use on agricultural lands. Homeowners, on the other hand, have access to a wide and abundant array of pesticides, no regulations on their use or licensing requirements, and almost no opportunities for education about the potential impacts of their spraying.

One thing is clear: Pesticides have a disastrous effect on pollinators. Millions of pounds of pesticides are applied to farms, fields, lawns, flower beds, and roadsides every year. Insecticides kill pollinators directly, while herbicides reduce the diversity and abundance of the flowering plants that supply them with pollen and nectar. Many pesticides degrade slowly, lingering as a toxic hazard to pollinators and other wildlife.

How Pesticides Affect Bees

Foraging bees are poisoned by pesticides when they absorb the fast-acting toxins through the **integument** (the outer "skin" that forms the exoskeleton), drink toxin-tainted nectar, or eat pesticide-covered pollen or micro-encapsulated poisons. If they are foraging while pesticides are being applied, the spray or dust covers them, killing significant numbers of bees in the field. If they are foraging on recently sprayed fields, they absorb toxins from the residues on plants, which kill them more slowly. Even dormant ground-nesting bees and their larvae are not necessarily safe, as they succumb to soil fumigants applied to kill root pathogens in farm fields.

Smaller bees (the majority of North America's native bees) are more sensitive than honey bees. With a larger surface area relative to their body volume, they absorb doses that are relatively higher. They are killed by lower concentrations of poisons, and thus insecticide residues on plants remain at toxic levels longer for smaller bees. After a significant kill, beekeepers may find thousands of dead honey bees in and around each hive. One can only imagine the thousands of dead native bees that at the same time are scattered around the landscape out of sight.

New Threats

To further complicate these threats, a new class of systemic insecticides has been developed in recent years. These products, which mimic the toxins found in nicotine, are applied as seed treatments, foliar sprays, and root treatments. The chemicals are then absorbed and transported by the vascular system throughout the plant. Some research suggests that these chemicals may be sequestered in flower nectar or pollen, and that pollinators may be poisoned as a result. The threat is so significant that European countries have recently restricted the use of these chemicals after complaints by beekeepers. Butterfly caterpillars feeding on the leaves of host plants treated with these systemic products are also at risk. Scientists are even researching the threat these insecticides pose to leafcutter bees that harvest leaf pieces to line their nests.

Because of their low toxicity to mammals, these chemicals have been widely embraced by various agricultural and landscape industries. Increasingly, these chemicals are being used

SMALLER DOSES

Even less-than-lethal doses of pesticides can cause problems. Bees that are exposed outside the nest may have trouble navigating their way back after foraging, or they may be unable to fly at all. Other symptoms include aggressive or agitated behavior, jerky or wobbly movements, or paralysis, all of which make foraging and nest building difficult. Sub-lethal doses can result from direct contact with pesticides or from toxins brought into the nest with nectar and pollen. They may impair egg laying in the nest or harm the developing larvae, and impact the next generation of bees.

PART 2

Taking Action

TAKING ACTION for pollinators can assume many forms. Protecting, enhancing, or restoring wildflower-rich foraging habitat is the most significant step you can take for bees, butterflies, and other flower visitors. In addition, you can provide nest sites for native bees, host plants for butterflies, and overwintering refuge for other insects. When these basic habitat essentials are provided, you will want to manage them in ways that ensure long-term productivity. Finally, advocate for pollinators. Share their story with neighbors and policymakers. Let your own conservation efforts be an example, and use pollinator conservation as a framework to achieve multiple goals. Actions for pollinators will also support other wildlife, beautify our world, and support sustainable farms and urban environments.

5 Strategies to Help Pollinators

WHETHER YOU ARE WORKING in your backyard or the back forty, the conditions you create to support the greatest variety and abundance of pollinators will have all of the following features:

A diversity of plants, preferably native. A variety of plants with overlapping blooming times provides flowers for foraging throughout the seasons. Native flowers, which are adapted to local soils and climates, are usually the best sources of nectar and pollen for native pollinators. (See chapters 6 and 10 for plant recommendations.)

Nesting and egg-laying sites, with appropriate nesting materials. Bees require areas of untilled, unmulched, partially bare ground or woody vegetation, as well as nesting materials such as leaf pieces or mud. Butterflies need appropriate host plants for their caterpillars. The egg-laying needs of flies and beetles are not easily defined, but they can be provided by a diverse, pesticide-free habitat.

Sheltered, undisturbed places for hibernation and over-wintering. Many insects live for more than a year, but they are active adults for only the last few days or weeks of their lives. Pollinators require secure places during the months when they are dormant or hibernating.

A landscape free of poisonous chemicals. The use of pesticides in gardens, on farms, and in managed landscapes is a major threat to pollinators. Pesticides should be kept away from pollinator habitat.

△ *The basic needs of bees — flowers, nesting areas, hibernation sites, and no pesticides — can be easily provided in any location.*

The Human Landscape

Many bees and other pollinators can adapt to patchy, human-influenced landscapes because in natural conditions, the plants and nest sites they need are spread randomly and inconsistently across landscapes. Does this mean people should not be concerned because pollinators can adapt to whatever humans do? Sadly, no. Several butterfly species are known to have gone extinct in the United States due to human activities, and a large number of other pollinators (including many of Hawaii's 62 native bee species) are known to be vulnerable.

EXAMPLES OF WELCOMING HABITAT

STUDIES DONE in Western Europe show that the habitats that support the most diverse bee populations are old grasslands, heathlands, and hedgerows. All of these contain a diversity of flowers, dead wood, and areas of bare soil for nesting sites, as well as mud, resin, and other nest-building materials. Shady places, particularly conifer forests, provide fewer flowers; they are the least used by bees but may be important to moths and other insects.

△ *In many areas, agricultural or urban development has removed habitat from the landscape.*

CRITICAL MASS OF HABITAT

Although many pollinating insects can adapt to changing conditions and often rebound from the effects of natural changes in their environment — including fires, floods, droughts, and windstorms — it's impossible to protect a diversity of pollinators without a critical mass of diverse habitats. That critical mass is being lost in many landscapes.

Not only is there less natural habitat than ever before, but the land that surrounds it may also be inhospitable to pollinators because it lacks food plants or nesting sites. In rural areas, the fields created by large-scale agriculture are too big for some bee species to cross to reach forage or nests. Tilling can destroy shallow bee nests or block the emergence of bees deeper in the ground. The crops grown on many farms are wind-pollinated members of the Grass family (wheat, corn, rice, grass, barley, oats, and so on) and have no value as bee forage, while those crops that do offer nectar and pollen usually provide them only in a brief, though abundant, burst.

In urban areas, landscapes around developments tend to be dominated by easy-to-maintain lawns and shrubs chosen for their colorful foliage rather than for their flowers. When there are flowers, often they are nonnative plants or highly ornamental varieties that provide little or no food value for native pollinators.

You can improve these conditions where you live if you consciously plan, create, and maintain habitat patches — the bigger the better — in urban, suburban, and rural areas. We can plant a diversity and abundance of nectar- and pollen-laden native plants, we can provide nesting habitat, and we can choose not to use pesticides, or at least exercise extreme caution.

Four Steps to Success

Providing habitat for pollinators can be as simple as planting a small garden. It can be daunting, however, for people making decisions about how to provide or improve habitat in a larger landscape. To help develop a pollinator conservation strategy, landowners and managers can follow a four-step approach:

1. **Recognize existing pollinator habitat** that is already present.
2. **Protect that habitat** and avoid causing undue harm to the pollinators already present.
3. **Provide new habitat for pollinators** (which is the focus for most of the following chapters of this book).
4. **Manage land** in a way that maintains the habitat and minimizes disturbance to pollinators.

The first two steps require very little outlay of cash and a relatively small time commitment. The third step, developing habitat, requires more thought and effort. The details provided in this book will make this more-intensive third step easier for those interested in doing something to increase the number of native pollinators. The fourth step, site management, overlaps some with step 2 but addresses the management of large areas of habitat, such as parks or natural areas. It also addresses the potential hazards of different site and weed management tools such as mowing, burning, or grazing, and adjusting management plans to minimize these hazards.

For home gardeners or others creating habitat in small sites, burning and grazing are not likely options. Here it may be more a case of being sure to wield the hoe carefully and consider how to alter garden maintenance to reduce disturbance to pollinator habitat.

We can plant a diversity and abundance of nectar- and pollen-laden native plants, we can provide nesting habitat, and we can choose not to use pesticides, or at least exercise extreme caution.

▷ *In the aftermath of a fire, it appears that little has survived. But well-planned fires can be a key management tool to maintain rich and diverse plant and pollinator communities.*

winters naturally limit tree growth and thereby maintain open habitats. In others, there are areas where the vegetation is disturbed (for example, burned or mowed) frequently enough that forests, trees, and shrubs do not grow back and thick layers of thatch do not build up (which makes it harder for solitary bees to nest and native wildflowers to grow).

Historically, fire has been widely used to maintain open landscapes, and in many smaller sites, mowing or grazing keeps vegetation under control. While manicured gardens obviously do not require the use of such drastic management tools, some large areas of habitat do, in order to prevent takeover by trees.

As mentioned earlier, however, adult bees, butterflies, flies, and beetles all can be killed by fast-moving fires. Pupating moths and butterflies are also vulnerable to fires, as well as to mowing and grazing. If an entire habitat is burned or mowed all at once, land managers run the risk of removing many of the insects they have worked hard to cultivate on their land. Therefore, it is important to tailor these practices to protect the smaller remnants of habitat present today; see chapter 7 for guidance.

The chapters that follow detail how to enhance habitat for native bees by addressing the major constraints to populations of native bees: forage availability, nest site availability, overwintering sites, and pesticide use. Each chapter describes how to provide these habitat resources and how to manage them to protect the insects you are cultivating.

Providing Foraging Habitat 6

ONE EFFECTIVE WAY to increase local pollinator numbers is to increase the flowers available to them. The best way to do this is by cultivating a landscape that includes a diverse range of plants to provide pollen and nectar throughout the local growing season. Such habitat can take the form of designated pollinator meadows ("bee pastures"), butterfly gardens, hedgerows of flowering trees and shrubs, streamside and rangeland revegetation efforts, and even flowering cover crops or pollinator lawns.

- ✿ Consider starting small and expanding over time.

- ✿ Incorporate a succession of flowers in order to provide blooms throughout the entire growing season.

- ✿ Include several different flower species known to be good sources of pollen and nectar or butterfly host plants.

- ✿ Native plants are better.

- ✿ Design new habitats to be free of pesticides.

△ *The creation of flowering habitats such as this patch of native plants on a California farm, benefits from careful planning.*

▽ *Using volunteers for planting can help limit costs and, be a way to engage the local community.*

PLANNING AHEAD

Before you jump into a pollinator conservation project, first take some time to plan ahead. Consider your options for where to create your habitat, how it will be maintained over time, the size of your budget, and the potential sources of funding or volunteer help for larger habitat projects.

For backyard gardens, this is a fairly simple process of working with what you have and experimenting with different locations, plants, and nests. Weed management is typically straightforward and a normal part of gardening.

For farmers and natural area managers, such an effort requires more thought. It is critically important, for example, not only to choose a good site for the habitat, but also to conduct site preparation and thorough weed removal prior to planting. In this way, the final habitat may take longer to come together but will require less work in the future and have a significantly higher chance of success.

The Budget

Early in the planning stages of larger pollinator-habitat conservation projects, think about the costs of site preparation, plant establishment, and long-term maintenance. Your biggest expenses will be plant material and labor, especially for the site preparation necessary to remove weeds and weed seeds.

△ *Grants are available to assist with both the planning and the implementation of habitat projects.*

There are grants and cost-sharing programs available to support projects for habitat creation and protection, particularly for farmers, nonprofit organizations, educational institutions, and municipal agencies. Your local Cooperative Extension Service office, conservation district, local library, and the Internet can supply information on private foundations, corporations, federal and state agencies, city governments, and environmental organizations that support such projects.

Federal and state agencies, such as the U.S. Fish and Wildlife Service and state departments of fish and wildlife, offer grants to support management of at-risk or endangered species. The U.S. Department of Agriculture's Natural Resources Conservation Service administers several financial assistance programs that offer farmers incentive payments and technical support for pollinator habitat projects. These USDA conservation programs are discussed in more detail in chapter 11.

Planning for Public Sites

When planning a project for a public site such as a wildlife preserve or suburban park, it is important to involve local people and other interested parties early in the process. This may mean establishing citizen advisory groups or consulting local residents, but getting their commitment to the project can draw in valuable expertise — and result in fewer obstacles later. Volunteers may include people from the neighborhood, gardeners and farmers, and/or members of local environmental organizations.

This is also a good time to begin recruiting volunteers for habitat creation and upkeep tasks, as these can be expensive if done with hired labor. Boy Scout, Girl Scout, and Camp Fire USA groups, for example, are often interested in doing conservation and habitat maintenance work.

SITE SELECTION

In any project, finding the right location will probably involve a compromise between the ideal and the practical. You'll have to evaluate the existing conditions: the slope and aspect of the site, the location of trees, the size of the area you are working in, and its current uses. In many places there may be practical considerations, including how equipment and materials will access the site for maintenance tasks such as watering or weeding.

Begin by studying the site so that you know what it offers in plant communities, topography, and soil moisture. For large areas, use aerial photos or maps to create a diagram showing existing habitat features. For a garden, this can be as simple as sketching plans, literally, on the back of an envelope. If there are areas with good foraging plants or potential nesting sites, focus your efforts on improving or enlarging them. Then look at the land between these areas: if it has little habitat value, consider adding patches of flowering plants or nesting and egg-laying sites to create stepping stones or habitat corridors between and among the principal pollinator areas.

If this land is important for other wildlife, you will not want to disturb it in order to create a flower-rich foraging area for bees, and it may in any case be valuable for pollinators in

▽ *Things to consider when selecting a site for habitat improvement include soil and aspect, current vegetation, irrigation, and neighbors.*

◁ *Wetlands and other areas unsuitable for development are often excellent sites for flowers.*

other ways. Woodland edges that are too shady for flowers, for example, may contain beetle-riddled snags and rotting logs in which bees can nest or on which syrphid flies can lay eggs. Similarly, wetlands may not suit ground-nesting bees, but they may well contain flowers for foraging and probably twigs or reed stems that offer nesting sites for tunnel-nesting bees.

Chapters 10 through 14 provide ideas for creating pollinator habitat in a wide variety of specific settings. Below are several other general considerations to keep in mind when choosing a project site.

Habitat Size and Shape

Habitat patches that are bigger, rounder, and closer to other patches are generally better than those that are smaller, of uneven shapes, and more isolated from one another. These configurations increase the area-to-edge ratio, which may limit encroachment by weeds or negative effects of adjacent land management practices, such as insecticide applications.

Sun Exposure

The best pollinator habitat sites will also have good sun exposure. Many pollinating insects can actually generate heat to warm their flight muscles enough to fly, but they are small and

PLANTING UNUSABLE AREAS

Consider that some otherwise unusable areas can be perfectly suited for new pollinator plantings. For example, septic fields can be planted in pollinator-friendly wildflowers. Shrubs can be planted on slopes that are too steep to mow. Similarly, small retention ponds around parking lots can be converted to flowering rain gardens, incorporating pollinator plants such as some species of native rose, willows, or Joe-Pye weed.

△ *South-facing sites provide the warm conditions that many insects prefer.*

lose this heat quickly to the air. Pollinating insects also need to be able to see the sky in order to navigate. Ideal pollinator habitat areas should thus receive a lot of sun.

A southern exposure can provide the warmest habitat, but it isn't necessary. South-facing sites also may require using more drought-tolerant plants. The main characteristic to consider is that the site be in full sun for at least part of the day and that it be mostly open, with few large trees.

Soil Conditions

Soil type is also an important consideration when selecting a site. Along with fertility, factors that influence plant establishment include soilborne plant diseases, herbicide residue, drainage, salinity, pH, organic content, and compaction. Some of these factors can be determined through a soil test, one of the services often provided by the local Cooperative Extension Service. Such a test may help you decide what species to plant or if the soil needs amending.

Site Maintenance

During the project's planning stages, consider long-term management in order to minimize the effort needed for future maintenance. Choose plants that are appropriate for the local growing conditions and that require minimal care. Plant suggestions for various regions of the country are provided in chapter 10. You can find the best information regarding the most appropriate plant materials by taking this book along to local nurseries, native plant societies, master gardener meetings, or your local Cooperative Extension Service office.

Habitat Design

Once you have determined the location, shape, and size of your pollinator habitat, you can focus on the specifics of the planting, such as plant selection, plant density, how plants are organized, and the inclusion of grasses for weed control and soil stabilization.

Planting Layout

Research suggests that flower groupings (clumps) of at least 3 feet (1 m) in diameter of an individual species are more attractive to pollinators than species that are widely and randomly dispersed in smaller clumps. Large clumps of individual species are easier for flying pollinators to find in the landscape, especially in the case of small urban habitats or small pollinators with flight ranges as short as 500 feet (152 m). For a natural look, these clumps can be distributed at random in a landscape, rather than in regularly spaced straight lines. In a large area of habitat, planting clumps may be impractical and not necessarily important so long as flowering plants are abundant.

Large, rounded planting blocks minimize the edge around the plantings and thus reduce susceptibility to invasion by weeds. Rounded blocks also blend better into many landscapes, although square or angular plant groupings may be well-suited to formal gardens. Linear corridor plantings (along a fence line, hedge, or road) are often more practical.

> **GROUP SPECIES FOR EFFICIENCY**
>
> Groupings of single flower species reduce the energy required for foraging because pollinators can spot the plant quickly (minimizing their search time). And with an abundance of flowers in one location, pollinators can move quickly and efficiently from flower to flower and collect a full load of pollen and nectar to bring back to the nest.

Flight Range

The flight range of pollinators is a necessary consideration for restoration and management of pollinator habitat. The distance a pollinator can fly varies among species, and thus the distance between food and nesting sources must be carefully considered. This may be most important for bees because — unlike butterflies, flies, and beetles — they transport pollen and nectar to a nest and therefore are locked into visiting the

▽ *Small bees such as this sweat bee (genus* Halictus) *fly less far than large ones.*

flowers surrounding their nest. Other pollinators may forage much more widely, roaming across the landscape in search of food or egg-laying sites, sometimes over many miles — even hundreds of miles, as in the case of monarch butterflies.

The ideal is to have nesting and forage resources in the same area. Bees are able to adapt to landscapes in which nesting and forage resources are separated, as long as these two key habitat components are not too far apart.

The distance a bee can fly between nest site and forage area is related to its size. Small species may fly no more than 500 feet (152 m) while larger species such as bumble bees may fly more than a mile (1.6 kilometers). A general rule of thumb is to have flowers no more than a few hundred feet (100 m) from potential nesting areas. With bumble bees, however, blueberry and cranberry farmers sometimes seek to establish habitat some distance from the field. In more heavily managed farm landscapes, you should work to have a patchwork of blooming plants, including flowering crops and wild plants on field margins or in-field insectary plantings, separated by no more than 500 feet (152 m).

△ *To support crop pollination, habitat needs to be within a few hundred yards of the farm field.*

The quality and size of habitat patches also affects pollinator populations. If habitat patches are too small to support pollinator populations and the distance between patches is greater than the foraging range of those species, the populations may suffer declines or the pollinator community composition may change. On the other hand, pollinators can survive in fragmented habitat if suitable nesting sites are available and if foraging sources are located within their flight range.

Plant Diversity

Diversity is a critical factor in the design of pollinator plantings. Natural flower-rich habitats may have 50 or 100 species, but for most conservation areas, as few as 10 carefully chosen plant species will provide a good foundation. Gordon Frankie, a professor at the University of California in Berkeley, and his

△ *Studies show that the best bee habitat contains a diversity of native plants.*

students have found that when eight or more species of plants are grouped together at a single site, they tend to attract a significantly greater abundance and diversity of bee species. From this foundation, a richer habitat can develop with subsequent plantings, or by colonization from nearby natural areas.

In some studies, pollinator diversity continues to rise with increasing plant diversity and starts to level out only when 20 or more different flower species occur at a single site. With several plant species flowering at once, and a sequence of plants flowering through the growing season, habitats can support a wide range of pollinators. If there are particular pollinators you wish to support, identifying and planting their key host plants may increase their abundance.

The size of the pollinators, the length of their tongues, and the accessibility of nectar or pollen within the flower are all factors in determining the floral preferences of diverse pollinators. Including a diversity of plants with different flower sizes, shapes, and colors, as well as varying plant heights and growth habits, will support the greatest numbers and diversity of pollinators.

Diverse plantings that resemble natural native plant communities are the most likely to resist pest, disease, and weed epidemics. Plant species found in association with each other in local natural areas are likely to have the same light, moisture, and nutrient needs and are more likely to thrive when grouped together. Thus, it is very useful to look around at natural areas in your community to see what plants are growing together, and which of these flowers seem to have the most visitors.

Many species of bees can feed from flowers such as asters that have a simple structure and easily accessible nectar. Some bees (notably some bumble bees) have long tongues or are big enough to push petals aside; they sometimes prefer more complex flowers such as lupines and salvias in which the nectar is hidden deep inside, often with a higher concentration of sugar.

Flies have short mouthparts best suited to open or small flowers such as yarrow, fennel, alyssum, or asters. Butterflies, moths, and hummingbirds can probe deep tubular flowers to reach nectar with their long tongues. Beetles often feed at blooms, such as goldenrod, where the nectar or pollen is easily accessible and there is plenty of perching space.

Bloom Time Succession

To provide a continuous food supply, choose at least three different pollinator plants within each of the three blooming periods: spring, summer, and fall. Under this plan, at least nine blooming plants should be established in pollinator enhancement sites; more is even better.

INCLUDE EARLY AND LATE BLOOMERS

It is especially important to plant flowers that bloom in the very early spring. These flowers are a critical resource for early emerging bees such as bumble bee queens, mining bees, and mason bees. An abundance of early to mid- spring blooming flowers also will help jumpstart populations of the handful of native bee species — such as bumble bees and some sweat bees — that produce multiple generations each year. These bees can be very abundant.

Adequate forage available early in the season will increase reproductive success and lead to more bees in the middle and end of the year. Early forage also may encourage bumble bee queens that are emerging from hibernation to start their nests nearby, or simply increase the success rate of nearby nests. It is equally important to include plants that flower late in the season to ensure that queen bumble bees are strong and numerous going into winter hibernation.

△ *Willow (genus* Salix) *is an early bloomer that can nourish emerging bees.*

Importance of Grasses

Large wildflower plantings should include at least one native warm-season bunch grass or sedge adapted to the site. Grasses and sedges provide host plants for some butterflies. They also supply potential nesting sites for colonies of bumble bees and overwintering sites for many insects. The combination of grasses and herbaceous (nonwoody) plants also forms a tight living mass that will resist weed colonization. And grasses are essential to produce conditions suitable for burning, if that is part of the long-term management plan.

Take care to keep grasses from taking over pollinator sites. Choose grass species that grow actively in the summer (also known as warm season grasses), when these are appropriate for your area. Anecdotal evidence suggests that tall grasses crowd out herbaceous plants more easily than do short grasses, and that grass species that grow actively in the spring (cool season grasses) often outcompete wildflowers.

If you are starting your pollinator habitat from seed, the amount of grass should not exceed 30 percent of the overall seed mixture. Planting in the fall, rather than spring, favors wildflower development over grasses.

△ *Early blooming flowers, such as camas lily* (Camassia qua-mash), *provide an important boost to spring bees.*

△ *California poppy (Eschscholzia californica)*

Native plants are four times more likely than nonnative plants to attract native bees, and native plant genera support three times as many species of butterflies and moths as introduced plants do.

△ *Transplants are more expensive than seed but establish and produce flowers more quickly.*

Whether you are transforming a lawn into a flower-rich meadow, planting a bee pasture to support crop pollination, or adding forage plants to an existing natural area, you need to find appropriate nursery stock or seed. Transplants will more likely survive drought and competition from weeds in the first year and provide habitat sooner than seed will. Some flowers, such as purple coneflower, take two years to bloom from seed, but transplants usually bloom the first year.

Seed, on the other hand, costs much less than live plants and may make economic sense if you are working on a large scale — landscaping a new park, farm, or business campus, for instance. And some species, such as poppies, do not transplant well but are easy to grow from seed; sowing is preferable for these regardless of the scale. For uncommon wildflowers and similar plants that are not in high demand, seed may be the only available source.

An alternate approach is to plant a combination of live transplants and seed, so that some plants bloom during the first year. This allows a wider range of plant selection and brings down the cost.

Native Plants

Native plants are generally the best choice for native pollinator habitat. Gordon Frankie's research demonstrates that native plants are four times more likely than nonnative plants to attract native bees. The presence of native vegetation significantly increases the abundance of butterflies and moths. Research by Douglas Tallamy and his colleagues at the University of Delaware has demonstrated that native plant genera support three times as many species of butterflies and moths as introduced plants. Even more startling, native woody plants used as ornamentals in gardens supported 14 times as many species as introduced ornamental species.

Native plants offer a number of additional advantages. In general, they:

* Do not require fertilizers
* Require fewer pesticides, if any, for maintenance
* Require less water than nonnative plantings
* Provide permanent shelter and food for other wildlife
* Are less likely to become invasive than nonnative plants
* Promote local native biological diversity

For an example of a native plant mixture designed specifically for pollinators, see Sample Native Seed Mix for a Pollinator Meadow on pages 120–121. While the specific plants in this seed mix may not be appropriate for your area, the general characteristics of the mixture (diverse colors, sizes, bloom times) are applicable to meadow-type plantings across much of North America. Use it as a starting point, along with the recommended plants featured beginning on page 272, to develop your own similar plant mixture.

Seed Sources

Where available and economical, native plants and seed should be procured from local ecotype providers. **Ecotypes** are locally adapted genetic populations within a species. **Local ecotype** refers to seed and plant stock harvested from a local source. Locally sourced plants generally establish and grow well because they are adapted to the local climatic conditions. Shopping for local ecotypes can help build markets for these plant materials and, as a result, help maintain or increase the genetic diversity of nursery stock. Start with the suppliers and wildflower organizations listed in the appendix; they can help you locate suppliers for your area. If purchasing in large volumes, look for seed certified for purity and viability; in the United States, state crop improvement agencies perform the certification.

Transplants

For most gardens and small landscape areas, you probably will not be growing all your pollinator plants from seed, but instead will be using live plants in the form of **plugs** (small potted plants grown in plastic trays), or larger potted plants. In

---

PROBLEMS WITH COMMON ORNAMENTALS

Some common ornamental plants, including ornamental varieties of viburnum, pansies, and double-petaled sunflowers, often do not provide sufficient pollen or nectar to support pollinator populations. Other common ornamentals such as butterfly bush (*Buddleja*) may become invasive and colonize natural areas at the expense of diverse native plant communities.

△ *Good site preparation — especially weed control — is important for good plant establishment.*

will inhibit planting operations, while leaving behind a bed that is firmly packed and ready to accept your plants or seed. Seeds and rhizomes of perennial weeds in particular need to be eliminated. Annual and biennial weed seed may still be present, but with several years of proper management (such as mowing annual weeds before they flower in the first year after planting), even large seeded areas can become fairly free of these weeds.

Because seeds are much more susceptible to weed competition than transplants are, site preparation prior to seeding should be performed in one of three basic ways:

* Without cultivation by smothering or solarizing or by using broad-spectrum herbicides
* With cultivation through aggressive tillage or sod removal
* A combination of tillage and herbicides

The appropriate method will depend on the size of the planting area, whether herbicides can be tolerated (for example, is the site an organic farm or garden?), equipment availability, and the current abundance of weeds or weed seed at the site.

METHOD 1: SITE PREPARATION WITHOUT CULTIVATION

Site preparation without cultivation is relatively low-cost, controls soil erosion, prevents dormant weed seeds in the soil from rising to the surface where they may germinate, and does not

---

ADVANCE WEEDING

Most soils are full of dormant weed seeds waiting for enough light and water to germinate. Prior to planting, your goal should be to eliminate as many of these competitors as possible.

require tillage equipment. The method is particularly useful for large areas and those where difficult weeds are growing.

For organic farms and gardens, smother existing vegetation with newspaper or cardboard, or solarize the area with plastic. Generally, the ground cover will need to be weighted down and left in place for at least a year before all the vegetation under it is dead. Use UV-stabilized plastic drop cloths for very large areas.

In larger seedbeds, existing vegetation can be killed with nonpersistent broad-spectrum herbicides (for an organic alternative, consider horticultural vinegar). This is a low-cost approach that can be extremely effective with minimal labor. For maximum efficiency, apply these treatments when weeds are actively growing. Repeated treatments are often required, especially for larger plantings. For example, early spring applications may not kill warm-weather weeds that grow actively or germinate in the summer. In some cases, a full season of monthly treatments is required to eliminate all existing weeds. In all cases, follow the instructions on the product label.

METHOD 2: **SITE PREPARATION USING CULTIVATION**

For areas too large to smother with paper or plastic, and where the use of herbicides is objectionable, existing vegetation can be removed with aggressive, repeated tilling. Break up turf grass before tillage; remove it by hand or with a walk-behind sod cutter, a machine that often can be rented from local hardware stores.

Site preparation using only cultivation may be preferred by organic gardeners, and by farmers with tillage equipment. This, of course, is a time-consuming and labor-intensive process, especially for areas that contain difficult perennial weeds. Where perennial weeds are prevalent, a planting site might require a full season of tilling

---

USING A ROTOTILLER

To create a weed-free seedbed with a rototiller, follow these steps.

- In spring, till as deeply as possible (up to 6 inches [15 cm]) to break up the soil and pull up perennial weeds. Smooth the surface with a rake or lawn roller.
- Once a month for the entire growing season, till the soil as shallowly as possible (roughly an inch [2.5 cm] deep) to kill germinating weeds but keep new weed seeds from being churned to the surface. Irrigate the area several times a month to encourage weed seeds to germinate.
- By midautumn, most weed seed should be depleted. Perennial wildflowers can be sown in mid- to late autumn, and annual wildflowers the next spring.

repeated every two weeks to eliminate all the dormant weed seed and rhizomes.

After the existing vegetation is gone, cultivated soils need to be smoothed and lightly tamped prior to planting. This is an easy task to do in a small patch with a soil rake or a turf roller. For larger areas, a cultipacker, roller, or similar tractor-drawn equipment will be necessary to finish seedbed preparation.

METHOD 3: SITE PREPARATION USING A COMBINATION OF HERBICIDES & TILLAGE

An extremely clean seedbed can be produced through a combination of very shallow cultivation and herbicide applications in alternating months through an entire growing season. Most site preparation and planting techniques, such as tilling, cultivation, or mulching, will destroy the habitat nest sites of ground-nesting bees. This is unavoidable and over time, if the tilled sites are left undisturbed, they will be recolonized.

Seeding Rates

For small areas, seed packages will often provide recommended seeding rates. For larger areas, however, calculating the proper seeding rates for native plants can be a complex task. In many cases, the seed vendor will provide a recommendation of how many pounds or ounces of a single species are needed to plant a certain area.

A better way to understand seeding rates is to think about seeds per square foot. Many native plant specialists, including people who restore prairies, recommend seeding rates for wildflowers and other herbaceous plants in the range of 20 to 40 seeds per square foot. (See pages 120–121.)

△ *Even small machinery like this ATV-towed broadcaster makes seeding of large sites easy.*

To translate this rate into a total amount of seeds to buy, ask the seed supplier how many seeds are in a given weight, usually an ounce or pound. Many native seed suppliers automatically include this information in their catalog descriptions. Then determine the size of your planting area in square feet. Using these two numbers, you should be able to estimate the amount of seed to purchase.

Sowing Seed

Successful plant establishment from seed can be challenging, but for large areas, and where particular transplants are not available, it is the only way to go. Unlike annual garden seeds, most native perennial plants are best established by dormant planting during late fall or even during the winter. Seeds of many perennial plants require exposure to cold temperatures and damp conditions before germination can occur.

For small garden areas, simply plant seed according to the package instructions. For larger areas, however, you have several planting techniques to consider: scattering seed by hand (broadcasting), frost-seeding, using a mechanical broadcaster, or using a no-till native seed drill.

METHOD 1: SCATTERING SEED BY HAND

Broadcasting seed by hand is a low-tech, low-cost option that can be very effective on bare soil. For even distribution, small seeds should be mixed with a slightly damp inert material such as sawdust, peat moss, or vermiculite.

When broadcast seeding, divide the mixture into two equal quantities. Apply one half of it to the planting area in one direction, and then work in a perpendicular direction to spread the second half of the mixture. This will result in fairly equal seed distribution throughout the field.

To prevent seed from being washed away by heavy rain or eaten by birds, protect the soil surface with a thin (approximately half-inch [12 mm] thick) layer of weed-free straw

△ *Hand sowing may be necessary on steep slopes.*

mulch. On smaller plantings, consider using a **floating row cover** (light fabric) instead.

METHOD 2: FROST-SEEDING

In cold climates, an alternative to broadcast seeding directly onto the soil surface is to "frost-seed" an area by broadcasting over snow-covered soil on a sunny, late-winter day. In this method, the seed warms in the sun, works its way through the melting snow, and gradually deposits itself on the soil surface. The melting snow helps settle the seed, ensuring good seed-soil contact and providing the moisture needed for germination.

Frost seeding works best if the seedbed is prepared before it snows, so that the seeds settle onto bare soil. The method also works for seeding small-seeded herbaceous plants (such as clover) into snow-covered areas of short mowed or grazed grass.

METHOD 3: USING A MECHANICAL BROADCASTER

For seeding larger areas of bare soil, a mechanical broadcaster will make the job easier. Options include a seed spreader mounted on an all-terrain vehicle, or a tractor-mounted drop seeder. The latter is typically used for alfalfa and grass mixtures. The standard seedbox agitators can be replaced with special native seed bristle agitators. Follow the equipment manufacturer's instructions for installation.

△ *Specialized drills for native seeds are available in many areas.*

As with hand scattering, divide the seed mixture into two equal quantities, and plant each half in separate perpendicular passes to ensure even distribution. Protect the newly broadcast seed with a half-inch (12.7 mm) layer of weed-free straw mulch.

METHOD 4: USING A NO-TILL NATIVE SEED DRILL

Specialized native seed planters are available in many areas, particularly the Midwest where prairie restoration efforts take place on a large scale. These devices are typically drills that plant seed in rough, untilled soil, and through the stubble of herbicide-killed vegetation. This is advantageous in areas with a large amount of dormant weed seed that may germinate when brought to the surface by cultivation. Seed drills also provide ideal planting depth and spacing; hence, they require half the amount of seed that broadcasting requires for the same area (roughly 15 to 20 seeds per square foot).

The drawback of drill seeding is equipment availability. In some locations, these machines can be rented, typically from private conservation or local hunting organizations, as well as some state departments of natural resources. Some large native seed producers also have these machines and will provide custom planting on a contract basis.

ESTABLISHING POLLINATOR HABITAT FROM TRANSPLANTS

Herbaceous transplants can be installed in bare, cultivated soil or in cleared areas within existing vegetation. Large areas can be cleared of existing vegetation using the same methods described above for preparing sites for seeds. Site preparation for transplants should focus on the removal of unwanted, competitive weeds adjacent to the planting holes, and amending soil around the planting holes as necessary to improve its texture.

Planting Herbaceous Transplants

The process of installing container-grown plants is very straightforward. Dig a hole the same depth as the container, remove the plant from the container, and position the point where the roots meet the stem slightly below the surrounding soil surface. Fill the hole surrounding the plant, making sure to cover the potting mix completely with the native soil to prevent moisture from wicking out of the potting mixture. First-year transplants benefit from a layer of straw or leaf mulch after going dormant in the fall. Keep the mulch a couple of inches away from the main stem.

If you can't plant container stock right away, be sure to keep it in the shade and water often — small pots dry out quickly. Most container-grown native plants can be transplanted throughout the growing season, even when the plant is dormant (has died back to the ground). If you can't time your planting to take advantage of a season with abundant rainfall, plan on keeping it well-watered for several weeks after planting.

△ *On smaller sites, transplants may be best.*

Site Preparation for Trees and Shrubs

In general, hedgerows or plantings of woody species require significantly less site preparation than herbaceous plants. Remove unwanted existing woody species with weed wrenches, machetes, chainsaws, branch loppers, or brush-hogs. After being cut back to ground level, many woody plant species will sprout new trunks. For permanent removal, these sprouts will require repeated cutting until the stored energy reserves in the root system are depleted; stumps also can be treated with a stump herbicide. Use mulch or weed mats to control herbaceous weeds.

Planting Trees and Shrubs

Trees and shrubs can be planted whenever the nursery stock is available and the soil can be worked. Planting depth is critical: woody plants should be planted at or slightly above the depth at which they were grown at the nursery. (Look for the soil line where stems meet roots.) Prune off any broken or obviously damaged roots, and take care not to damage the roots further during the installation process.

△ Trees should be planted no deeper than the depth at which they were growing in the pots.

After planting, mulch around the base of trees and shrubs to help maintain soil moisture and suppress competing weeds. Most newly planted trees and shrubs do best without staking. Trunk movement caused by wind helps develop proper trunk taper and resilient root systems. Animal browsing can be a problem for small woody plants. Where this occurs, trunks can be protected with various types of trunk guards, such as a purchased trunk guard or a collar of hardware cloth.

MAINTAINING NEW PLANTS

No matter what or where you plant, post-planting maintenance is vital. The most time-consuming task — watering newly installed plantings — tapers off quickly as the habitat matures. Natural nutrient cycling in native plant ecosystems eliminates the need for supplemental fertilizers. Typically, the only maintenance necessary after the first season is the regular monitoring and removal of weeds.

Watering

Once established, native plants typically do not need supplemental irrigation. Newly installed woody or herbacaeous transplants usually require routine watering the first year; the exact amount depends on the species, location, precipitation, and time of year. Water every couple of weeks if it doesn't rain; occasional deep soakings are better than frequent shallow watering. The original plant vendor or a local nursery often can provide the necessary guidance on local water requirements.

Seed germination and seedling survival also benefit from regular irrigation during the first few months after planting. With or without supplemental water, many native perennial seeds germinate slowly; they may even take several seasons to germinate. Although this may seem to be a drawback, this trait allows for the identification (and removal) of faster-growing weeds in the seedbed.

△ A drip irrigation hose is an efficient way to provide water to a newly planted hedgerow.

Weed Control

If properly mulched, new transplants stay well protected from weed encroachment during the first season or two. Any weeds around transplants usually can be controlled by hand-weeding alone. Avoid herbicides, to prevent damage to your newly planted pollinator plants. Weed eradication takes time, but diligence and persistence pay off with fewer weeds each year so that pollinator plants can enjoy better growing conditions.

Large areas seeded with native perennials need more protection from annual and biennial weeds during the first couple of years. (All perennial weeds should be destroyed prior to planting.) The best way to control annual and biennial weeds is by regular mowing of the seedbed during the first growing season. Plan on mowing every 4 to 6 weeks during the first two growing seasons, or whenever many annual weeds appear ready to flower. Mow at whatever height is necessary to remove those weed flower heads. This will prevent most weeds from flowering, while leaving the slower growing native perennials untouched.

Although this seems like a very slow process, the pace benefits perennials, which typically do not flower in the first year. Instead, they will be building strong root systems, and by the third year have far less competition from annual weeds.

Spot treatments for weeds in seeded plots include applying herbicide with a sprayer or wiping with an herbicide wick (typically a herbicide-soaked sponge on a handle). This is particularly useful for isolated clumps of pernicious weeds such as Canada thistle. Weeds should not be pulled if at all possible, especially during the first year when the surrounding seedlings are still developing their root systems and deeply buried weed seeds may surface, to germinate and compete.

If perennial weeds were destroyed prior to planting, and annual weeds mowed during the first year of growth, the majority of the remaining weeds in the second year should be biennial species, such as Queen Anne's lace and common burdock. These plants develop a thick taproot during their first year, and then flower during the second year before dying. Rather than mowing the area at regular intervals, wait until most of the biennial weeds are about to flower, then mow only as necessary at a height that cuts off most of the weedy flower heads.

By following these steps, large seeded areas should require only periodic spot treatment of weeds after the first two years of establishment. Depending upon location, large areas of habitat also may require ongoing maintenance to prevent encroachment of trees and woody shrubs. Mowing (ideally during the dormant season) continues to be one option for removing woody plants. Burning, grazing, and spot treating with herbicides are other options. See chapter 7 for greater detail on these long-term management tools.

OTHER POLLINATOR FOOD SOURCES

Creating the type of wildflower habitat described above will support a great diversity and abundance of bees, butterflies, flower flies, beetles, and hummingbirds. However, nectar and pollen are not the only foods sought out by some of these pollinators. Some species of adult butterflies, for example, also get sugars from overripe fruit. In certain parts of the country, such as the Midwest and Southeast, consider placing cut pieces of fruit — oranges, bananas, and peaches are especially appealing — in the garden for butterflies to find.

Muddy puddles, animal carcasses, and dung also provide vital nutrients that nectar alone cannot. You probably do not want to scavenge for fresh roadkill for your garden, but damp patches of sand and muddy streambanks supply some of these same benefits and may be worthy of protecting.

▽ *Echo azure (Celastrina echo) butterflies drinking at wet soil.*

SAMPLE NATIVE SEED MIX
FOR A POLLINATOR MEADOW

This seed mixture is for a typical Midwestern pollinator meadow. For other regions, the plant species will change but the overall design criteria remain the same:

- The mix includes a minimum of three blooming plants for each season (spring, summer, and fall).
- It includes flowers of different shapes and colors.
- Host plants for butterflies of special local interest are included (milkweed for monarchs, lupine for Karner blues).
- Low-growing, warm-season bunch grasses are added for bumble bee nests.
- The total number of seeds in this mix, divided by the number of square feet per acre (43,560), results in 21 seeds per square foot. This amount is within the target range of 20 to 40 seeds per square foot.

Wildflowers

COMMON NAME	SCIENTIFIC NAME	SEEDS/SQ. FT.
Wild lupine	Lupinus perennis	0.014
Smooth penstemon	Penstemon digitalis	2.64
Ohio spiderwort	Tradescantia ohiensis	0.37
Wild bergamot	Monarda fistulosa	2.69
Purple prairie clover	Dalea purpurea	2.75
Pale purple coneflower	Echinacea pallida	0.69
Culver's root	Veronicastrum virginicum	0.34
Butterfly milkweed	Asclepias tuberosa	0.24
Prairie blazing star	Liatris pycnostachya	0.99
Purple giant hyssop	Agastache scrophulariifolia	2.71
New England aster	Symphyotrichum novae-angliae	3.49
Giant sunflower	Helianthus giganteus	1.03
Showy goldenrod	Solidago speciosa	2.41
		Total Seeds: 20.36

Bunch Grasses

COMMON NAME	SCIENTIFIC NAME	SEEDS/SQ. FT
Little bluestem	Schizachyrium scoparium	0.16
Prairie dropseed	Sporobolus heterolepis	0.06
		Total Seeds: 0.22

Veronicastrum virginicum

Dalea purpurea

Asclepias tuberosa

Liatris pycnostachya

SEEDS/OZ.	SEED OZ./ACRE	SEEDS/ACRE	BLOOM PERIOD	FLOWER COLOR
1,000	0.6	600	Early	Blue
115,000	1.0	115,000	Early	White
8,000	2.0	16,000	Early	Blue
78,000	1.5	117,000	Mid	Blue
20,000	6.0	120,000	Mid	Purple
5,000	6.0	30,000	Mid	Purple
75,000	0.2	15,000	Mid	White
3,500	3.0	10,500	Mid	Orange
10,750	4.0	43,000	Mid	Lavender
59,000	2.0	118,000	Mid	White
76,000	2.0	152,000	Late	Purple
15,000	3.0	45,000	Late	Yellow
105,000	1.0	105,000	Late	Yellow
Total Pounds: 2.1 lbs.		**Total Seeds: 887,100**		

SEEDS/LB.	SEED LBS./AC.	SEEDS/ACRE		
8,800	0.8	7,040		
14,000	0.2	2,800		
Total Pounds: 1.0 lbs.		**Total Seeds: 9,840**		

7 Reducing Impact of Land Management Practices on Pollinators

* When land is managed well, pollinators abound. When it is managed carelessly, pollinators suffer.

* Maintain open, early successional habitat.

* Conduct mowing and burning when plants and pollinators are dormant, and limit disturbance to only one-third or one-fourth of the landscape each year.

* Adopt integrated pest management strategies to reduce the need for pesticides.

TO MAXIMIZE THE NUMBER of pollinators, it is essential to maintain the diverse native plant communities and nesting habitats that support them. In some cases, however, the practices needed to support a diversity of pollinator plants — such as prescribed burning or mowing — may be detrimental to the pollinators themselves. In other cases, areas maintained for pollinators may be exposed to pesticides used to control crop pests, invasive species, or perceived public health threats such as mosquitoes. The key is in understanding how to balance these factors.

Pollinators are important in almost all North American ecosystems, but the greatest abundance of pollinators occurs in open landscapes with plentiful herbaceous plants. Prairies, meadows, and shrub-dominated landscapes all can provide habitat for pollinators. Even some forested areas such as California's oak woodlands can provide many of the resources that pollinators need. As stated earlier in this book, it is important to identify the best pollinator habitat in the landscape you are managing.

Landscapes dominated by herbaceous plants tend to be temporary. These areas appear early in the long process of plant succession, during which open areas gradually transition through low-growing vegetation to shrubs and eventually, in many regions, to forests.

Historically, there were sufficient areas in various stages of succession to support habitat for a wide variety of animals. Today, however, landscapes are dominated by intensively managed land through which habitat is scattered in fragments. Loss of grasslands and meadows has taken place across the United States. Few tallgrass prairies remain in the East and Midwest, and prairies and meadows in the West have not fared much better. These areas are now a fraction of their former size and the animals that lived on

△ *Landscapes dominated by farming leave only scattered fragments of habitat.*

them have been relegated to smaller and smaller areas. Since many of these areas are so small and fragmented, they are more likely to be overrun by invasive plants and native species (such as trees) that march across the landscape and change these areas from grass- and wildflower-dominated prairie to areas dominated by shrubs and trees.

So what to do? These areas need to be managed or they will no longer provide viable habitat. But the very management practices used could lead to the extirpation or extinction of some of the species that rely on these habitats. Some level of disturbance is usually required to limit woody plant growth

Reducing Impact of Land Management Practices on Pollinators ✽ 123 and to control invasive weeds. Too much disturbance can be just as detrimental as too little. This chapter details how to balance these competing interests.

GRAZING

Grazing in natural areas and rangelands is a common practice. Livestock grazing greatly alters the structure, diversity, and growth habits of vegetation, which in turn affects the associated insect community.

Negative Impacts on Pollinators

If not managed appropriately, the ecological impact of grazing can be severe. Grazing during periods when flowers are already scarce may result in insufficient forage available for pollinators such as bumble bees, which forage as late as November in some areas, or even year-round in warm climates. Grazing during spring when butterfly larvae are active on host plants can result in larval mortality or reduce pollinator populations by reducing their food sources.

Other harmful effects of grazing on pollinator habitat include destruction of potential nest sites, destruction of existing nests, and direct trampling of adult bees. Studies of the effects of livestock grazing on bee populations suggest that increased intensity of livestock grazing reduces the number of pollinator species present.

SHEEP & BUMBLE BEES

A study done at San Francisco State University found that uncontrolled sheep grazing in mountain meadows in the Sierra Nevada removed enough flowering plants to eliminate bumble bees from some study sites.

▽ *Sheep may appear benign, but their grazing has caused the loss of bumble bees in some meadows.*

Although only limited research has been done on the impacts of grazing on pollinators in the United States, there is a considerable body of work from other countries. In Argentina, researchers compared insect communities in grazed and ungrazed areas and found that insect diversity, abundance, richness, and biomass were all lowest in intensively grazed areas. A study of four different grazing regimes in Germany found that as grazing intensity increased, the diversity and abundance of invertebrates, including butterflies and ground-nesting bees, decreased. A similar study in Switzerland found old fallow fields that had not been grazed harbored many more rare and specialist species of butterflies than managed pastures or recently fallowed land, most likely due to the reduction of nectar resources in grazed pastures.

Even light grazing can cause declines in some species. A study that directly addressed light grazing as a method of preventing grassland from reverting to forest found that it significantly reduced the numbers of bog fritillary butterfly (*Boloria eunomia*) on site. This effect is largely attributable to changes in vegetation structure, loss of preferred forage sources, and a decline of the host plant in grazed plots.

One study in Arizona focused explicitly on the impacts of domestic livestock grazing on invertebrate communities in an area that had not been grazed historically. The results clearly show that the invertebrate species richness, abundance, and diversity were all greater in the ungrazed sites. The author suggested that since insects in the Southwest had not evolved in the presence of buffalo or other large grazing animals, adaptations to grazing pressure had not developed, making the insects more sensitive to the modern presence of cattle.

Positive Aspects of Grazing

Grazing is not necessarily harmful in all instances. Many parts of the world have experienced grazing pressure from both domesticated and wild animals for millennia, and the indigenous flora and fauna have adapted. Even in areas where grazing has recently been introduced, light levels of rotational grazing can help maintain an open, herbaceous plant community that is capable of supporting a wide diversity of butterflies and other pollinators.

Some research suggests that grazing can be beneficial for certain insect communities, especially by managing invasive plants and succession. Grazing does need to be carefully planned and implemented, however, to be effective. A Swiss

△ *Heavy grazing results in degraded vegetation communities and trampling of insect nests and larvae.*

study found that while grazing was an effective management tool for limiting the incursion of woody plants, responses to grazing varied greatly among butterfly species. The authors suggest that any management regime be attentive to historical and species-specific characteristics of the site. They recommend using a diversity of management techniques on a regional scale in order to preserve the greatest diversity of insect pollinator habitat. Grazing is usually only beneficial at low to moderate levels when the site is grazed for a short period followed by ample recovery time — and when the grazing has been planned to suit the local site conditions. The best approach where grazing must occur is to rotate livestock onto only one-third or one-fourth of the management area in a single season, leaving ungrazed areas to provide reservoirs for pollinator populations.

MOWING

Mowing often replaces grazing on small sites, on sites where topography permits equipment access, or on sites such as suburban parks where livestock may be inappropriate. Like grazing, mowing can suppress growth of woody vegetation, but it can also negatively impact insects through direct mortality, particularly for egg and larval stages that can't avoid the mower. Mowing also creates a sward of uniform height, destroying features such as grass tussocks that provide structural diversity to the habitat and offer potential nesting sites for bumble bees and other pollinating insects. In addition, mowing abruptly removes almost all flowers.

The differences between an ultimately beneficial mowing regime and a detrimental one are timing, technique, and scale. Avoid cutting flowers in the treated area by mowing during nonblooming seasons — when the plants have died back or are dormant, such as in fall or winter. An exception to this is a weed management program where there is a narrow window of opportunity for optimal control of the target species.

△ *Where grazing is not viable, mowing may be the only practical alternative.*

There are mowing techniques that will benefit pollinators as well as other wildlife. Many of these were developed to protect ground-nesting birds, but they also contribute to reducing impacts on pollinating insects.

* **Use a flushing bar** when mowing, to encourage animals to move on before the blades reach them.
* **Cut at reduced speeds** — less than 8 mph (12 km per hour) — in order to give wildlife more time to move out of the way.
* **A high minimum cutting height** of 12 to 16 inches (30–40 cm) leaves a greater depth of vegetation within which insects (of all life stages) can remain.
* **Avoid mowing at night.** Even insects sleep, so cutting during the day when they are more active means they are more likely to flee.
* **Finally, to minimize these effects** and allow sufficient space and time for pollinator populations to recover, mow in a mosaic (or patches) instead of an entire site.

If weed management is the short-term objective, however, such as in newly planted perennial patches, it may be necessary to mow more than once a year. In this case, try to limit mowing to patches of weeds.

FIRE

Fire has played an important role in many native ecosystems, and controlled burns are an increasingly common management tool. Effects of fire management on insect communities are highly variable. If used appropriately, fire benefits many pollinators through the restoration and maintenance of suitable habitat. If it is too frequent, widespread, or intense, however, it will eliminate numerous invertebrates.

In Midwestern prairie systems, the use of fire as a management tool is based on the supposition that prairie plant species are adapted

▽ *Prescribed fire is used to maintain flower-rich prairies.*

to wildfires, and thus can cope with regular burns. Although this may generally work for plants, it is true for pollinators only if there are adequate unburned areas nearby to provide sources of colonizers that may return to the burned habitat. In small, more isolated habitat fragments, prescribed burning can have much more deleterious effects on the pollinators because there may be no local populations to recolonize the site. One study found that overall species diversity decreased in burned prairie sites, as well as the abundance of all but one of the species measured. The results suggest that burning a small habitat fragment in its entirety could risk extirpating some species because of limited recolonization from adjacent habitat.

Proper Burning Protocol

Prescribed burning should be designed so a single fire does not burn an entire area of pollinator habitat. A variety of approaches can help.

Follow a program of rotational burning in which small sections — 30 percent of a site or less — are burned every few years. This will ensure adequate colonization potential and refuge habitat for insects.

Leave skips — small unburned patches — intact as potential microrefuges as a fire moves through an area.

Avoid burning an area too frequently. Based on a variety of studies, it appears that five to ten years between burns allows adequate recovery of pollinator populations, depending on the ecosystem and specific management goals.

Avoid high-intensity (hot) fires unless the objective for a prescribed fire is brush or tree removal (for example, controlling pinyon-juniper, buckthorn, or mesquite). Lower-temperature prescribed burns conducted early or late in the day, or from late fall to early spring, are not only preferable for pollinator survival but also reduce impacts on other wildlife species such as reptiles and ground-nesting birds.

ALTERNATIVES TO PESTICIDES

There are many things you can do to reduce, eliminate, or limit pesticide use. A plant that is growing vigorously, with minimal stress, can often avoid or outgrow many diseases and insect pests without the help of pesticide. Plant health starts with choosing species appropriate to local conditions.

It is also important to recognize and work with naturally occurring pest controls. A healthy and diverse landscape has the necessary habitat to encourage native predators and/or parasitoids of insect pests. Pesticides may eliminate the natural enemies of pests and cause chronic problems. Fortunately, many of the same practices that support pollinators will support these other native beneficial insects, further reducing the need for pest control.

△ *Pesticides not only have the potential to harm pollinators but can also eliminate other beneficial insects, such as those that prey upon crop pests.*

In gardens, practices such as handpicking and crushing larger insects or spraying a stream of soapy water to dislodge smaller plant pests may be practical and effective. Good sanitation practices — removing infected leaves and the previous year's crop debris from the area — also limit the spread of disease.

in urban areas than on agricultural lands. Use of agricultural chemicals is restricted to licensed applicators; home gardeners, on the other hand, can buy any available product and use it without training or supervision. As a result, chemicals are often applied in backyards in quantities far greater than those recommended by the manufacturer.

Pesticide Drift

Insecticide use on large farms and large managed landscapes has its own unique problems. Pesticides may be applied in a variety of ways, from backpack sprayers to crop dusters. Pesticide drift from ground or aerial spraying onto adjacent crops or wildflowers may kill foraging bees close to the source, but drift can continue to cause damage for a mile and a half or more.

△ *Pesticides can drift hundreds of yards from their target.*

Factors affecting the extent of drift include weather, application method, equipment settings, and spray formulation. Weather-related drift increases with temperature, wind velocity, convection air currents, and during temperature inversions. Low humidity and high temperature conditions promote drift through the evaporation of spray droplets and the corresponding reduction of particle size, which allows them to float in the air for much longer.

Even a light wind can cause considerable drift. Pesticide labels will occasionally provide specific guidelines on acceptable wind velocities for a particular product. Wind-related drift can be minimized by spraying during early morning or in the evening when wind velocity is often lowest. Counterintuitively, some wind (2 to 9 mph) helps reduce drift by creating turbulence that helps spray droplets fall out of the air more quickly.

Midday spraying is also less desirable because as the ground warms, rising air can lift the spray particles in vertical convection currents. These droplets may remain aloft for some time and can travel many miles.

Similarly, during temperature inversions spray droplets become trapped in a cool lower air mass and move laterally above the ground. Inversions often occur when cool night temperatures follow high day temperatures; they are often characterized by foggy conditions.

REDUCING DRIFT POTENTIAL

Optimal spray conditions for reducing drift occur when the air is slightly unstable with a very mild steady wind (2 to 9 mph). Ideally, temperatures should be moderate and the air should have some humidity.

Spray application methods and equipment settings also strongly influence the potential for drift. Follow these guidelines to minimize impact.

Avoid aerial spraying and mist blowers whenever possible, since small droplets are most likely to drift long distances.

Operate standard boom sprayers at the lowest effective pressure and with the nozzles set as low as possible.

Select nozzles capable of operating at low pressures (15 to 30 psi) to produce larger, heavier droplets that will deliver the insecticide within the crop canopy where it is less likely to be carried by wind currents.

Calibrate sprayers regularly, regardless of the chemical or type of application equipment used, to ensure that excess amounts of pesticide are not applied.

CASE STUDY
Agrecol Corporation, Wisconsin

VISITORS TO THE FARM of the Agrecol Corporation in southern Wisconsin are often astounded at what they see — hundreds of acres of native prairie plants growing as row crops. When the big fields of plants such as lance-leaved coreopsis (*Coreopsis lanceolata*) or purple prairie clover (*Dalea purpurea*) are in bloom, it is an amazing sight.

"Our goal is to produce native seed for habitat restoration," says Agrecol president Mark Doudlah. "Part of that habitat restoration mission is a commitment to native pollinators. For us, as seed producers, it's a simple equation: no bees equals no seed."

Agrecol, whose name combines "agriculture" and "ecology," presents a unique business model, synthesizing native plant production with modern tools such as center-pivot irriga-tion and combine harvesters. Most of the seed Agrecol produces ends up in prairie restoration projects or USDA-administered farm conservation programs.

Agrecol's own on-farm pollination manage-ment consists of a multispecies approach, utiliz-ing managed honey bees and leafcutter bees as well as habitat conservation for wild bees. Habitat conservation includes reduced tillage in areas with high populations of ground-nesting bees, limited mowing to protect bumble bee nest sites, nighttime pesticide applications to prevent bee poisoning, and an aggressive IPM scouting program that ensures insecticides are only applied when absolutely necessary. The end result is a state-of-the-art farm with a rich diversity of beneficial native insects. ✿

Nesting and Egg-Laying Sites for Pollinators

8

POLLINATOR POPULATIONS benefit most from flower-rich, pesticide-free foraging areas if suitable egg-laying or nesting sites are nearby. Addressing this need may not entail much extra work, because by providing good foraging habitat you likely are already supplying suitable egg-laying sites for many pollinators, particularly butterflies, as well as undisturbed areas/ground with potential nest sites for bees. Still, with extra attention, you can significantly increase nesting opportunities.

❀ Provide nesting and egg-laying sites for a variety of pollinator species.

❀ Clean and replace artificial nests regularly.

❀ Don't move native bees or previously used nest materials outside of their native ranges.

❀ Leave some bare, unmulched ground.

the morning sun to fall on the entrance holes. Direct sunlight later in the day can make nests too hot, causing the developing brood to die.

Nest site success — persuading bees to move in — is greatest when blocks are firmly attached to a large visible landmark to help bees locate their nests when returning from foraging. In open landscapes such as farms or prairies, this might be a building, a nest-block shelter painted in contrasting colors, or an isolated tree. This is especially important on farms where great numbers of bees are desired. In smaller sites — an urban garden, suburban park, or a business campus — affixing nests to a building is also good, but there may be other suitable locations such as a fence or a firmly installed stake. In these sites, as well as natural areas where fruit or crop production is not the primary goal, careful placement with an obvious landmark offers the best chance of occupancy.

△ Whether stuffed inside a previously used plastic bucket or nestled in a purpose-made sleeve, hollow stems make effective nests. ▽

Mounting the Block

It is important to ensure firm mounting of the block so that it does not move or shake in the wind, because movement disrupts nesting and larval development. The actual height from the ground does not matter much, although if the nest is too low (less than a couple of feet [0.6m]), rain splash may dampen it and vegetation may cover it. Blocks that are too low are also cooler than is optimal.

STEM BUNDLES

In addition to wooden blocks, artificial nests can be constructed with bundles of reed, teasel, cup plant (*Silphium perfoliatum*), or sections of bamboo garden stakes cut so that a natural node forms the back wall of the tunnel.

Setting Up a Stem Bundle

Place the bundles in a sheltered location (such as the side of a barn or garden shed) with the stems horizontal to the ground — or pointing slightly down — and the holes facing east to get the morning sun.

DRIED MUD BRICKS

In deserts or semiarid areas, solitary bees will nest in cracks or cavities in soft sandstone, and in dry, exposed soil embankments. Some species, such as *Anthophora abrupta* and *A. urbana*, wet the hard soil surface with water or nectar to soften it enough to excavate tunnels. These species are quite common in the southeastern and southwestern United States, respectively; they are both important visitors of some fruit and vegetable crops.

To attract these species, dried mud bricks can serve as the equivalent of a wooden nest block. Such adobe bricks can sometimes be purchased, in which case you can increase their attractiveness to bees by drilling nesting holes following the size recommendations listed above for wood blocks.

Mounting a Mud Block

Mount the brick, or a stack of bricks, facing east or southeast to catch the morning sun. Adobe will not hold up well in wet climates, so the block may need sheltering from rain. Some commercially available adobe bricks have a stabilizing plasticizer mixed in; the plasticizer is not necessary and may not even be desirable.

CELL PARTITIONS AND LINING MATERIALS

Bees that occupy nest tunnels in wood or adobe construct intricate walls that partition off short lengths of the tunnel into a series of separate chambers. The leafcutters go one step further and construct a series of capsules made of layered leaf pieces that encase the developing brood. They use these additional nest materials to seal off and/or hide the nest entrances, protecting the brood from predators.

Depending on the species, bees construct these walls of mud, plant resins, chewed-up pieces of leaves or petals, fine pebbles, sawdust, or even cellophane-like glandular secretions. Bees have to collect the materials (except for the sawdust and glandular secretions) and carry them to the nest. Chances are these materials are already present in your area, but if you provide a diversity of native plants and some mud puddles on clay soils, you can guarantee the bees will find what they need.

Keep in mind that bees may not fill the entire length of a tunnel with cells, or they may die before an entire length of a cavity is filled. For these reasons, it can be difficult to tell if a nest tunnel is occupied just by looking for sealed entrances.

TUNNEL-NEST MAINTENANCE

Whether your nests are wooden blocks, twig bundles, or other materials, in order to be sustainable they will need routine management and regular replacement. This is important to prevent the buildup of bee parasites and diseases that affect the developing brood.

Chalkbrood

The hardest of these to control is the pathogenic fungal disease chalkbrood (*Ascosphaera* spp.). Several species of these fungi exist among cavity-nesting bees, all of which are different from the chalkbrood disease that attacks honey bees. Bee larvae become infested with disease spores through contaminated pollen, either collected from a flower by the mother bee, or accidentally spread when the mother bee emerges from a contaminated nest cavity.

After they are ingested, the chalkbrood spores germinate inside the gut of the developing larva, producing long filaments (**hyphae**) that eventually penetrate the gut wall and kill the larva. These cadavers pose a hazard to bees deeper within the nest block: to emerge from the nest they must climb over or chew through the spore-infested cadaver or cell. Bees that emerge under these circumstances have a high likelihood of spreading the spores to their own offspring. Similarly, bees searching for new unoccupied nest tunnels in which to lay their eggs frequently investigate, and often select, previously used tunnels. Over time, chalkbrood spores are spread throughout a nest block in this way.

Pollen Mites

Along with chalkbrood, pollen mites in the genus *Chaetodactylus* can be a persistent problem in nest blocks that are in continuous use for several seasons. Unlike the mites that attack honey bees, pollen mites do not feed on the **hemolymph** (blood) of the bee. Instead, pollen mites are cleptoparasites that feed on the pollen provision, causing the developing bee larva to starve.

Adult pollen mites are clearly visible only when viewed with a hand lens; they are usually white, tan, or orange in color and measure about half a millimeter in width. As with chalkbrood, adult bees may accidentally pick up mites at flowers while foraging, or from climbing through or investigating contaminated nest cavities. The migratory nymphal stage pollen mites cling to a bee's hairs and are transported to the nest, where they feed on the pollen provision and reproduce rapidly. In a single provisioned cell, mite numbers can quickly climb into the thousands over the course of the summer, after the nest has been sealed.

> **Routine management and regular replacement will prevent the buildup of bee parasites and diseases that affect the developing brood.**

FIGHTING PARASITES AND DISEASE

To reduce parasite and disease problems, try one of three approaches:

DISINFECT NEST BLOCKS

Line the tunnels of wooden nest blocks with tight-fitting removable paper straws. At the end of the nesting season (autumn), gently remove the straws (with bees in them) and place them in a ventilated container stored either in a refrigerator or in an unheated barn or garage.

The USDA Bee Biology and Systematics Laboratory in Logan, Utah, recommends disinfecting nest blocks after the straws have been removed by submerging them in a 1:3 bleach-water solution for a few minutes, ensuring that no air bubbles remain in the holes.

In the spring, place the old straws inside an emergence chamber, described in the next paragraph, and place new straws inside the disinfected block.

RENEW BLOCKS AND BUNDLES

Every two years, phase out wooden and adobe nest blocks and stem bundles by placing them inside an **emergence chamber** at the beginning of the season when the adults are active. This is a dark container such as a lightproof wooden box, a dark-colored plastic bucket with a tight-fitting lid, or even a sealed milk carton that has been spray-painted black to reduce light infiltration.

Drill a single 3/8-inch (10 mm) exit hole in the bottom of the light-proof container, and set the entire contraption outside, adjacent to a new nest block or stem bundle. As bees emerge from the old nest, they are attracted to the light of the exit hole and emerge to find the new nest hanging nearby. Locating the escape hole on the bottom of the dark container means the bees can crawl out, rather than attempting to fly out. At the end of the season, remove the old nest block or stem bundle from the emergence box and dispose of it.

Under this timeline, not all bees may emerge because some are **parsivoltine**, which means that they may skip a year before emerging. If this is a concern, leave nest blocks in the emergence box for two seasons.

OFFER SMALLER BLOCKS AND BUNDLES AROUND PROPERTY

Create multiple small nest blocks or stem bundles with only a few nest tunnels (four to six), and scatter them across your property, at least 25 feet (8.0 m) or more apart. This prevents the unnaturally high populations of bees found at nest blocks with many holes, and it mimics natural conditions of limited, spatially separated nest sites. These smaller nests also decompose more rapidly. They can simply deteriorate naturally as you add new small nests to the landscape periodically. Build these small nest blocks out of 8-inch (20 cm) lengths of 2 X 4s (5 cm X 10 cm), with holes drilled into the end, or use small bundles of bamboo or reeds.

Although pollen mites can't break through cell partitions, they can persist for many months without food, until a bee deeper within the nest emerges from the tunnel and breaks the partition walls, allowing them to escape. Mites often attach to the emerging bee, sometimes covering the bee in such numbers as to prevent it from being able to fly.

Providing Nests for Bumble Bees

Unlike solitary tunnel-nesting bees, which can be very particular about tunnel diameters, bumble bees are flexible in their nesting needs. They are looking for a warm, dry, roughly shoebox-size cavity. You can provide potential nesting sites for bumble bees by allowing grassy areas to grow dense and tall and then fall over, creating cavities in which bumble bees can find old rodent nests. This is most effective along the edges of wood lots and hedgerows.

You also can create dense piles of brush or fieldstones. These piles need not be large, but should be constructed in a way that provides dark, protected cavities sheltered from rain. In some regions, blueberry and cranberry farmers have even used old mattresses with holes poked in them to expose the stuffing as a way to potentially attract nesting bumble bees. Artificial nests also attract bumble bees, but occupancy is typically low — often far less than 25 percent.

LOCATING AND INSTALLING THE NEST

It is not unusual for bumble bees to nest under a deck, shed, or concrete stoop and never cause problems, but it is a good idea to locate nests away from places where people will be active. Avoid unnecessary disturbance, and be cautious when close to a nest. This does not mean you should avoid putting nests in your garden or never go near them, but put them more than 10 feet from paths, play areas, patios, or similar places, to be safe.

The best time to install a bumble bee nest box is in late winter or very early spring, when the first queens have emerged from hibernation and are searching for a nest site; in many areas, this is when the willows first start blooming. Once the nest is in place, be patient. It may take weeks for a bee to move in.

BUMBLE BEE NATURE

Most bumble bees are very gentle creatures, especially when visiting flowers. A few species can be aggressive in defending their nests, in particular *Bombus fervidus* and *B. pensylvanicus* in the East and *B. pensylvanicus sonorus* in the Southwest, all fairly common species.

Risks of Commercial Shipment

As mason beekeeping has increased in popularity, important questions are emerging about the risks associated with the nationwide shipment of blue orchard bees by commercial producers. There are two distinct subspecies of the blue orchard bee: *Osmia lignaria propinqua* in western North America, and *Osmia lignaria lignaria* in the East. The species is largely absent from the Great Plains region.

Most commercial producers of the blue orchard bee are located in the Pacific Northwest, where they rear the western subspecies. The bees these producers raise are marketed nationwide, resulting in the frequent shipment of *O. l. propinqua* to the eastern United States. The potential ecological consequences of the western blue orchard bee hybridizing with its eastern subspecies are unknown.

Of greater concern is the mass rearing of any single species (monoculture) and how it may simultaneously promote explosions in populations of pests and diseases. Shipments of blue orchard or other bees across North America have the potential to introduce locally nonnative parasites and diseases. The result can be catastrophic for wild local bees, and — sadly — we will likely never know that diseases have spread and solitary bees have been affected because so few people are tracking populations of these insects. Keep in mind that diseases and pests can build up in your nest blocks and can spread to other bees in the area.

For the consumer, there is another significant drawback — the western bees may not develop in sync with local eastern conditions, resulting in poor establishment and poor performance as pollinators. The net result is that people purchasing blue orchard bees over the Internet or through garden catalogs may be doing harm to their local pollinator populations.

△ *Purchased bumble bee colonies used for crop pollination should not be moved outside their native range.*

Commercial Bumble Bees

Similarly, the bumble bee industry, which caters to farmers and gardeners wanting live bumble bee colonies for crop pollination, currently produces only a single eastern bumble bee species, *Bombus impatiens*. Like the blue orchard bee, this species is shipped nationwide, often far beyond its native range. Many bee scientists now believe that these commercially produced bumble bees are responsible for the introduction of one or more diseases that have decimated several once-common bumble bee species across the United States and Canada.

Artificial Beds for Alkali Bees

The alkali bee (*Nomia melanderi*) is a ground-nesting species native to arid regions west of the Rocky Mountains. Growers of livestock forage crops first recognized the alkali bee's affinity for alfalfa blossoms in the 1940s. Since the development of management techniques for aggregations of alkali bee nest sites in the late 1950s, growers using the alkali bee have produced astounding yields of more than 2,000 pounds of cleaned seed per acre — this in the normally unproductive desert regions of the Great Basin.

In the wild, the alkali bee excavates nests in the crusted-over soils of alkali flats. In these areas, the arid climate combined with a high water table and the capillary action of the soil draws water to the soil surface. On the surface, salt crystals and minerals are deposited as the water evaporates, forming a thin crust that further limits evaporation and maintains a moist nesting environment in the otherwise arid region of the bee's native range.

Under optimal conditions, a single cubic foot of these alkaline soils may contain up to 50 nests, each with a dozen or more provisioned cells. Within these **playas** (or salt flats), individual bee nests are usually located in the upper 8 inches (20 cm) of soil and consist of tunnel entrances with a series of brood chambers located below the surface.

FINDING LOCAL ALTERNATIVES

There is an enormous need for managed agricultural pollinators in this country, but regional bee industries must be developed. It is very important to avoid importing bees that are not local, especially near wild areas, even if they are represented as the same species as local bees. Seek out commercial sources of bees that are native to your region, ideally local ecotypes, and talk with your local nurseries about supporting local mason bee industries. Alternatively, use already widely established non-native species such as honey bees or alfalfa leafcutter bees as managed pollinators.

▽ *An alkali bee bed can support millions of bees, nesting at densities of several hundred per square yard.*

△ *Alkali bee (Nomia melanderi)*
entering nest

Because they typically nest in flat alkaline desert soils, these bees' usefulness as pollinators is confined to arid locations. At the same time, they are susceptible to habitat loss from grazing and off-road vehicle use that is now common in the arid west. Managed alkali bees are increasingly uncommon in parts of their original range. However, artificially constructed soil beds continue to provide some habitat in the alfalfa seed–producing regions of the Great Basin and Pacific Northwest.

Alkali bees may find the artificial bed if other natural or managed nest sites are located within a mile or so. Otherwise, the bed may have to be stocked with bees, either adults or larvae, by cutting blocks of undisturbed soil from established beds and inserting them into the new one. This is normally performed during the winter when the larvae are inactive.

There is little published information regarding optimal bed size for managed alkali bee populations. Beds of approximately an acre (0.4 hectare) in size may produce thousands or even

HOW TO
CREATE AN ALKALI BEE BED

BUILD THE ARTIFICIAL BED in silty loam soils with no more than 7 percent clay.

HOW THE BED IS SET UP

Excavate a basin, usually 2 to 3 feet deep (0.6 to 1 m), and up to several thousand square feet (1,000 sq m). Line it with an impermeable layer (pond liner, plastic sheeting, or bentonite clay). On top of this, add a layer of gravel. Radiating outward from vertical standpipes, set a system of horizontal perforated drain pipes into this layer of gravel, with the vertical standpipes positioned every 50 feet (20 m). Finally, refill the basin with lightly compacted soil. Slightly elevate the soil level at the bed's center so rainwater does not pool on top.

To create the alkali crust, water is added to the standpipes. Moisture spreads through the perforated drainpipe and then rises to the surface to create a firm layer of salt that seals in soil moisture and slows further evaporation of water from the bed.

MANAGEMENT IS KEY

Maintaining the correct soil moisture where bees are actively nesting is critical. Water is added as needed through the standpipes. If excessive soil drying becomes a problem during the nesting season, bee beds can be shaded with either nursery shade cloth or military surplus camouflage netting suspended above the bed on permanent posts.

millions of bees. Since alkali bees may forage up to 5 miles (8 km) or more, they provide pollination services for multiple farms. In some areas, growers' cooperatives manage community beds — productive bee nest sites for many decades.

LARVAL HOST PLANTS FOR BUTTERFLIES

The idea of growing a plant so that a caterpillar can eat it might seem odd, especially to some gardeners. But in the big picture of creating healthy habitat, it makes perfect sense. Some butterfly and moth caterpillars are highly specialized in their food needs. Providing their larval host plants is an essential part of creating healthy habitat, given that it is during the caterpillar stage that these insects do almost all of their growing.

A monarch caterpillar, for example, may be only 1/16 inch (1.5 mm) long when newly hatched; within two or three weeks it can grow to 2 inches (50 mm) long, a 30-fold increase. Its increase in volume is far greater — approximately 10,000 times. The plants on which the females of different butterfly species will lay their eggs vary enormously. The monarch has a very restricted range of larval hosts, laying her eggs only on milkweed plants; the gray hairstreak, at the other extreme, is known to lay eggs on dozens of different plants from seven plant families.

△ *Showy milkweed* (Asclepias speciosa)

To provide the appropriate host plants, first you must know what butterflies are likely in your local area, and then match their larval requirements with local plant species. Start with the most common species to have the greatest chance of success.

You may discover that some very common butterflies are typically absent from certain areas. For example, monarch butterflies are rare or absent in most areas of the Pacific Northwest west of the

MORE ABOUT BUTTERFLY HOST PLANTS

PART 4 OF THIS BOOK lists larval host plants for common butterflies. It is not a definitive list; for plants specific to your region contact your local nursery, Cooperative Extension Service, or wildlife conservation organization, such as a native plant society, local North American Butterfly Association chapter, or Audubon chapter.

Alternatively, consult one of the regional butterfly guidebooks or websites listed in the appendix. The Butterflies and Moths of North America database (www.butterfliesandmoths.org) is one of the most comprehensive and can provide you with species lists for individual counties in the United States.

When considering any of these lists, remember that native plant species have been shown to support many more species of butterflies and moths than introduced species.

Cascade Range, so planting milkweed in that region is highly unlikely to attract them.

Nesting Sites for Flies and Beetles

There is a huge diversity of pollinating flies and beetles, and thus a huge diversity of foraging and nesting behaviors. Fortunately, the egg-laying and larval needs of these insects are likely to be addressed if you focus on managing habitat (food, nesting, and egg-laying sites) for bees and butterflies.

Forage for Flies

The foraging plants you provide for butterflies and bees will be home to spiders and aphids, the preferred larval food, respectively, of small-headed flies and many syrphid flies. The larvae of some syrphid flies live in the bottom of bumble bee nests, and the larvae of many bee flies are parasites of ground-nesting solitary bees, so providing for these bees takes care of other pollinators in the process. By tolerating some dead wood and leaving a few untidy corners, you will meet most of the egg-laying needs of the pollinator flies and the nutritional needs of their larvae.

Beetle Habitat

Most of the habitat requirements for pollinating beetles can be met as you manage landscapes for bees and butterflies. The life cycle of blister beetles is linked directly to that of bees. Tumbling flower beetle larvae develop in plant stems. The larvae of soldier beetles and soft-winged flower beetles hunt for prey in soil, leaf litter, and moss. Sap beetle larvae develop in all types of rotting materials; scarab larvae require decaying wood, usually stumps or fallen logs. The larvae of jewel beetles and long-horned beetles usually live in the wood of some trees.

If you have the little extra space it takes, and you know that your trees are not infected with diseases, or invasive, nonnative — and potentially devastating — pests such as the emerald ash borer or Asian long-horned beetles, think twice before removing snags and dying trees. These can provide benefits both for native pollinating beetles and for populations of wood-nesting bees that construct their nests within the abandoned beetle tunnels.

Pupation and Overwintering Sites

9

POLLINATORS, like most insects, have different habitat requirements, depending upon their life stage and the time of year. Earlier chapters discussed the needs of pollinating insects when they are active — as larvae and adults. This chapter focuses on the times when they are inactive — the egg and the pupa. Pollinating insects also may undergo **diapause** (a condition similar to hibernation) when conditions are inhospitable to activity.

* Pollinators need safe places of refuge when they are dormant or undergoing metamorphosis.

* Some insects overwinter in their immature stages; others overwinter as adults.

* Most bees and wasps overwinter in their natal nests, except bumble bees and social wasps.

* The monarch butterfly is the only pollinator in North America that migrates to warmer climates.

Diapause can occur when a species is at any of its life stages: egg, larva, pupa, or adult. Often, however, an inactive life stage coincides with the inhospitable season, and the insect passes through a harsh time of year as egg or pupa. In some species, for example with the bumble bees and some butterflies, diapause occurs during the adult state. Diapause often coincides with changes in temperature, rainfall or the lack thereof, and the availability of flowers and other food sources. For each of the inactive life stages or periods of inhospitable conditions, an insect requires a sheltered place.

BEES IN WINTER

For most bees, the nest serves both pupation and overwintering needs. Solitary bees spend most of the year in their brood cells, passing through the stages of egg, larva, and pupa protected either underground or in the shelter of a tree or twig. They emerge in the spring or summer to spend a few weeks active as adults. Some species of solitary bees, such as small carpenter bees, emerge at the end of the summer and overwinter as adults, usually in their original nests.

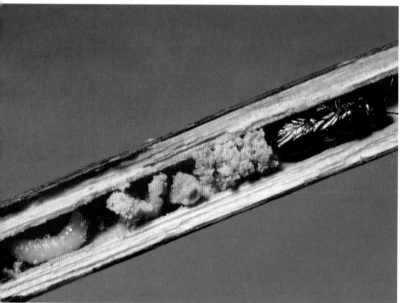

△ *Small carpenter bees (genus* Ceratina) *pass through egg, larva, and pupa stages before spending the winter as dormant adults in the nest.*

In contrast, most bumble bees die at the end of summer, leaving only the fertilized queens to hibernate. The queen bumble bee does not hibernate in her natal nest, which by then is likely to be moldy and maggot-infested; rather, she survives the winter by burrowing a couple of inches into soil or leaf litter. Preferred sites may be a north-facing bank or forest edge, a location shaded from the sun and covered with a thick layer of leaf litter or soil so that there is less fluctuation in winter temperatures. To protect themselves from freezing, hibernating bumble bees produce glycerol, a natural antifreeze.

Butterfly and Moth Pupae

The needs of Lepidoptera are slightly more complex than those of bees, as these insects may require separate places to pupate and to overwinter. Some species complete pupation in only a few weeks during the summer. Others overwinter as pupae, perhaps wrapped in cocoons or buried underground (in the case of many moths), or surrounded by a hardened chrysalis (as with butterflies).

Pupation normally occurs in tall grass, bushes, trees, fence posts, a pile of leaves or sticks, or the outside of a house. To find such a safe place, caterpillars may crawl many yards from their larval host plants. Keeping some untidy corners and piles of woody debris in your garden will give these creatures a place to go.

△ *Anise swallowtail (*Papilio zelicaon) *butterflies overwinter as a chrysalis.*

Protect Sites from Disturbance

Depending on the species, butterflies may survive the winter in any of the four life stages of egg, caterpillar, pupa, or adult. For those species that pass the winter as eggs or caterpillars, the best protection you can offer is to leave the areas where their larval host plants occur as undisturbed as possible during those months. Similarly, for species that overwinter as pupae, protecting potential pupation sites from disturbance during the cold months of the year is the best approach.

▽ *Burning of this meadow helps maintain an open habitat for butterflies. That activity must be balanced, however, with the need to protect overwintering and pupating insects.*

For areas of habitat that require management such as annual mowing or burning, late summer and fall is often a good time for these activities. It is often best to target management of these areas for times of the year when the dominant plant communities have senesced (and pollinators are dormant) because of drought or encroaching winter cold. In general, do what you can to manage one third or less of the site each year, and consider how you can protect some untidy areas from disturbance.

Provide Overwintering Sites for Adults

Although only a handful of butterflies such as anglewings and tortoiseshells overwinter as adults, many stores sell overwintering boxes for butterflies. Some are ornate and make attractive additions to a garden, but there is no evidence that the boxes work, at least not for butterflies. Spiders will move in, so the boxes do have some benefit for your garden. But there are more effective ways that you can help adult butterflies survive winter.

Under natural conditions, butterflies that overwinter as adults are likely to take shelter in tree cavities, under logs,

HOW TO
CREATE AN OVERWINTERING SITE

IT IS EASY TO CREATE an overwintering site by building a pile of logs or rocks. They should be constructed and placed to give shelter from prevailing winds and rain. Stack logs crisscrossed with gaps of at least 3 to 4 inches (8 to 10 cm). If you prefer, you can disguise a log pile by planting nectar and larval plants around it or a vine over it.

Rock piles are harder to make but may be less obtrusive. You don't have to use natural rock; this can be an opportunity to dispose of chunks of unwanted concrete. Vines, preferably native — though in a yard or garden, nonnative varieties will also work — growing over walls or buildings can offer good overwintering niches.

MAINTAINING THE SITE
The overwintering sites you create, like all pollinator habitat, require some maintenance. You may need to replace logs that have rotted away, or prune back vines. It's best to do these chores in the summer, so that you don't disturb the sites from late fall to early spring, when they may be occupied.

behind loose bark, under rocks, or within evergreen foliage. Human activities have inadvertently created other viable sites, such as stone walls, buildings, and fences.

Migration

Migrating species — in particular monarch butterflies, which overwinter in Mexico and coastal California — require foraging resources in their summer breeding sites and secure locations in which to overwinter. Just as important, though, is a corridor of nectar-rich habitat patches along their migration routes, across landscapes altered by agriculture and development.

Because monarch butterflies are very specific in their overwintering requirements, they are now limited to a few remnants of forest in the mountains of central Mexico (where more than 95 percent of them overwinter) and isolated groves of Monterey pine, Monterey cypress, and eucalyptus in California. The mild winters of the California coast are a perfect haven from the harsh cold weather in the interior of North America. Monarchs overwintering in California take advantage of this climate and use the same sites year after year. Congregations of overwintering monarchs gather at more than 200 sites along the California coast, from Mendocino County in the north to San Diego in the south. These places must be protected if the monarchs are to survive.

To sustain them on their journeys, monarchs require nectar for energy. On their return north in the spring, they also need native milkweed on which to lay their eggs. These needs can be met with stepping-stone habitat patches, which may be nothing more than the "weeds" growing in a field margin or alongside a road. A tolerant farmer or road maintenance crew who leaves these areas to grow rather than cutting or spraying them may help provide the feeding or egg-laying resources these migrating pollinators need.

△ Monarch butterflies pass the winter clustered in trees in areas with stable, low temperatures. This cluster is in a grove on the California coast.

10 Home, School, and Community Gardens

* Include native plants, heirloom flower varieties, and kitchen herbs in your garden.

* Have multiple plant species blooming throughout the growing season.

* Don't use pesticides.

* Protect existing natural nest sites, and enhance nesting opportunities.

A TYPICAL YARD fulfills many purposes. It's an attractive setting for a house, a place to relax on a summer afternoon, a play area for children, a setting for social gatherings. A typical school or community garden shares many of these goals, with the added opportunities to educate students and the public. Making it into a haven for pollinators is compatible with all of these uses.

Gardens and Flower Beds

You can design and plant flower beds with a range of native flowers and heirloom varieties that provide nectar and pollen for bees, or are the preferred caterpillar host plants for butterflies. In fact, designing gardens so there's something blooming at all times is a classic aesthetic objective that happens to benefit pollinators as well, when the correct plants are chosen. To round out an ornamental flower bed for pollinators, avoid using pesticides (especially systemic, which may persist for years), and include some patches of bare ground and bee nesting blocks. In larger gardens, there may be space for a flowery meadow, a small orchard, or a stone pile for overwintering insects. (See Part 4 for sample garden plans.)

△ *Even a small urban garden can support pollinators.*

If you maintain a fruit and vegetable garden, you probably already appreciate how essential pollinators are, and your vegetable plots can support them in large numbers. The fruits and vegetables you grow, from tomatoes to pumpkins and berries to apples, provide abundant flowers and are part of your pollinator habitat. Common herbs such as rosemary, lavender, mint, oregano, marjoram, and borage are excellent bee plants. If they are allowed to **bolt** (go to flower), unharvested garden plants such as radishes, broccoli, basil, and carrots will support bees and beneficial syrphid flies.

Community Gardens

Community gardens offer many opportunities to accommodate pollinator populations. For example, nesting blocks erected throughout the garden and flowers grown along the margins between plots will help support bees, as will keeping the garden free of pesticides. Hedge borders between the garden and adjacent parks or privately owned land can include shrubs such as rose, ninebark, currant, false indigo, and others. An unused garden plot could become a site for ground-nesting bees or, planted with appropriate flowers, serve as bee pasture. Cooperatively managed, it will benefit every gardener.

School Yards

Like residential yards and gardens, school grounds offer many opportunities for creating pollinator habitats. If your school already has a garden or wildlife area, it is easy to integrate the

---

PESTICIDES

Pesticide use is an important consideration in any garden designed for pollinators or butterflies. All of the flowers in the world will not help if the pollinators are then poisoned with insecticides. In fact, a beautiful pollinator habitat will concentrate the pollinators in one area, making a thoughtless pesticide application especially harmful. In most urban landscapes, these chemicals are simply not needed. See page 129 for an overview of alternatives to pesticides in a variety of landscapes.

LOW GROWERS AND LAWNMOWERS

SOME SCHOOL USERS might complain if the grounds appear untidy, but there are many low-growing flowering plants that can survive below the cutting height of a mower and provide good forage for pollinators. Plants that will thrive in such conditions include violets (*Viola*) and common selfheal (*Prunella vulgaris*), as well as nonnative bird's-foot trefoil (*Lotus corniculatus*) and clovers (*Trifolium*).

The addition of nesting sites and foraging areas will create an outdoor classroom that can be the source of stimulating biology and botany lessons and art classes, or just a fun place to play or relax during recess. An increasing number of classroom guides and curriculum materials are available that provide guidance on creating pollinator habitat and integrating it into school activities. In addition to these direct benefits, a pollinator habitat project that involves students, parents, local businesses, or other volunteers can help strengthen ties between the school and the community.

needs of pollinators by growing a range of suitable flowering plants and providing nest sites. But you can also provide foraging and nesting habitat by planting flowerbeds next to the buildings and along walkways, or by transforming the margins of sports fields into habitat corridors, with wildflowers planted adjacent to hedgerows and nesting sites in sunny corners.

As you include plants that are in bloom throughout the growing season, weight your school yard plantings toward species that flower early in the spring and late in the fall. Plants such as willow and wild indigo bloom early, and when combined with later-flowering goldenrods and asters, will provide more opportunities for students to observe the insects visiting these flowers while school is in session, rather than creating a learning lab that is only available to summer students.

▽ *Planning and planting a pollinator garden makes an excellent project that links school and community together.*

△ *Simple signs will identify pollinator habitat and inform the public.*

Garden Nest Sites

Chapter 8 provides a variety of recommendations for creating different nest types that can be used in garden settings. These structures can be entertaining, attractively designed, and educational. In addition to artificial nests, include some natural, untidy features that also provide nest sites. For ground-nesting bees, this may be semi-bare patches of soil in well-drained areas. Wood-nesting bees may be attracted to stumps, dead standing trees, or other plants with hollow stems. Bumble bees may nest in old rodent burrows, under thick grasses, in brush piles, or in stone walls. With creativity, these structures might be purposefully incorporated into a formal design plan.

Consider installing signs to educate visitors and students about the pollinator planting and nest sites. Signs adjacent to both natural and artificial nest structures provide a perfect opportunity to educate the public about the importance of pollinators and inspire visitors to create habitat at home.

Regional Plant Lists
for Native Pollinator Gardens

THE FOLLOWING plant lists include species that are native to their respective regions and commercially available from nurseries, specialty seed producers, and local native plant societies. The plants listed here tolerate a wide range of soil and light conditions. They are listed by season, to make it easier to design a garden that provides forage in spring, summer, and autumn. If particular species are difficult to find in your area, look for closely related ones. Part 4 includes more information on many of these plants, as well as additional plants to consider.

Spring-Blooming Herbaceous Plants

Wild lupine (*Lupinus perennis*)

Eastern waterleaf (*Hydrophyllum virginianum*)

Bicknell's cranesbill (*Geranium bicknellii*)

Summer-Blooming Herbaceous Plants

Smooth penstemon (*Penstemon digitalis*)

Wild bergamot (*Monarda fistulosa*)

Purple giant hyssop (*Agastache scrophulariifolia*)

Butterfly milkweed (*Asclepias tuberosa*)

Culver's root (*Veronicastrum virginicum*)

Joe-Pye weed (*Eupatorium fistulosum*)

Autumn-Blooming Herbaceous Plants

New England aster (*Symphyotrichum novae-angliae*)

New York aster (*Symphyotrichum novi-belgii*)

Canada goldenrod (*Solidago canadensis*)

Trees and Shrubs

American basswood (*Tilia americana*)

Common serviceberry (*Amelanchier arborea*)

Highbush blueberry (*Vaccinium corymbosum*)

Veronicastrum virginicum

Asclepias tuberosa

SOUTHEASTERN UNITED STATES

Spring-Blooming Herbaceous Plants

Spiderwort (*Tradescantia* spp.)

Eastern smooth beardtongue (*Penstemon laevigatus*)

Manyflower beardtongue (*Penstemon multiflorus*)

Spotted geranium (*Geranium maculatum*)

Summer-Blooming Herbaceous Plants

Virginia mountainmint (*Pycnanthemum virginianum*)

Summer farewell (*Dalea pinnata*)

Dense blazing star (*Liatris spicata*)

Spotted beebalm (*Monarda punctata*)

Annual blanketflower (*Gaillardia pulchella*)

Joe-Pye weed (*Eupatorium fistulosum*)

Autumn-Blooming Herbaceous Plants

Common sneezeweed (*Helenium autumnale*)

Pine barren goldenrod (*Solidago fistulosa*)

Giant ironweed (*Vernonia gigantea*)

Trees and Shrubs

Southern magnolia (*Magnolia grandiflora*)

Sourwood (*Oxydendrum arboreum*)

Carolina rose (*Rosa carolina*)

Smallflower blueberry (*Vaccinium virgatum*)

MIDWESTERN UNITED STATES

Spring-Blooming Herbaceous Plants

Smooth penstemon (*Penstemon digitalis*)

Wild lupine (*Lupinus perennis*)

Eastern waterleaf (*Hydrophyllum virginianum*)

Spotted geranium (*Geranium maculatum*)

Summer-Blooming Herbaceous Plants

Wild bergamot (*Monarda fistulosa*)

Purple giant hyssop (*Agastache scrophulariifolia*)

Butterfly milkweed (*Asclepias tuberosa*)

Purple prairie clover (*Dalea purpurea*)

Purple coneflower (*Echinacea purpurea*)

Prairie blazing star (*Liatris pycnostachya*)

Autumn-Blooming Herbaceous Plants

New England aster (*Symphyotrichum novae-angliae*)

Showy goldenrod (*Solidago speciosa*)

Riddell's goldenrod (*Solidago riddellii*)

Trees and Shrubs

Leadplant (*Amorpha canescens*)

Prairie rose (*Rosa arkansana*)

Pussy willow (*Salix discolor*)

American basswood (*Tilia americana*)

(lists continue next page)

Eupatorium fistulosum

Dalea purpurea

GREAT PLAINS AND PRAIRIE PROVINCES

Spring-Blooming Herbaceous Plants

White wild indigo (*Baptisia alba*)

Prairie spiderwort (*Tradescantia occidentalis*)

Largeflowered beardtongue (*Penstemon grandiflorus*)

Summer-Blooming Herbaceous Plants

Wild bergamot (*Monarda fistulosa*)

Rough (or tall) blazing star (*Liatris aspera*)

Showy milkweed (*Asclepias speciosa*)

Purple prairie clover (*Dalea purpurea*)

Narrowleaf coneflower (*Echinacea angustifolia*)

Compassplant (*Silphium laciniatum*)

Autumn-Blooming Herbaceous Plants

Smooth blue aster (*Symphyotrichum laeve*)

White heath aster (*Symphyotrichum ericoides*)

Showy goldenrod (*Solidago speciosa*)

Zigzag goldenrod (*Solidago flexicaulis*)

Trees and Shrubs

Prairie rose (*Rosa arkansana*)

Saskatoon serviceberry (*Amelanchier alnifolia*)

Leadplant (*Amorpha canescens*)

Chokecherry (*Prunus virginiana*)

Pussy willow (*Salix discolor*)

ROCKY MOUNTAIN REGION, INCLUDING CANADA

Spring-Blooming Herbaceous Plants

Rocky Mountain penstemon (*Penstemon strictus*)

Rocky Mountain beeplant (*Cleome serrulata*)

Perennial blanketflower (*Gaillardia aristata*)

Summer-Blooming Herbaceous Plants

Fireweed (*Chamerion angustifolium*)

Firecracker penstemon (*Penstemon eatonii*)

Purple prairie clover (*Dalea purpurea*)

Showy milkweed (*Asclepias speciosa*)

Autumn-Blooming Herbaceous Plants

Maximilian sunflower (*Helianthus maximiliani*)

Rocky Mountain goldenrod (*Solidago multiradiata*)

Smooth blue aster (*Symphyotrichum laeve*)

White heath aster (*Symphyotrichum ericoides*)

Trees and Shrubs

Yellow rabbitbrush (*Chrysothamnus viscidiflorus*)

Oceanspray (*Holodiscus discolor*)

Woods' rose (*Rosa woodsii*)

Saskatoon serviceberry (*Amelanchier alnifolia*)

Amelanchier alnifolia

Gaillardia aristata

CALIFORNIA AND THE SOUTHWEST

Spring-Blooming Herbaceous Plants

California poppy (*Eschscholzia californica*)

Lacy phacelia (*Phacelia tanacetifolia*)

Silvery lupine (*Lupinus argenteus*)

Summer-Blooming Herbaceous Plants

Showy milkweed (*Asclepias speciosa*)

Yellow beeplant (*Cleome lutea*)

Firecracker penstemon (*Penstemon eatonii*)

Wild buckwheat (*Eriogonum* spp.)

Autumn-Blooming Herbaceous Plants

Vinegarweed (*Trichostema* spp.)

Nevada goldenrod (*Solidago spectabilis*)

Eaton's aster (*Symphotrichum eatonii*)

Hayfield tarweed (*Hemizonia congesta*)

Trees and Shrubs

Mule–fat *(Baccharis salicifolia)*

Chamise *(Adenostoma fasciculatum)*

Mexican locust *(Robinia neomexicana)*

Rabbitbrush *(Chrysothamnus spp.)*

Redbud *(Cercis spp.)*

PACIFIC NORTHWEST AND BRITISH COLUMBIA

Spring-Blooming Herbaceous Plants

Lupine (*Lupinus* spp.)

Arrowleaf balsamroot (*Balsamorhiza sagittata*)

California poppy (*Eschscholzia californica*)

Perennial blanketflower (*Gaillardia aristata*)

Summer-Blooming Herbaceous Plants

Venus penstemon (*Penstemon venustus*)

Blue Mountain prairie clover (*Dalea ornata*)

Showy milkweed (*Asclepias speciosa*)

Wild buckwheat (*Eriogonum* spp.)

Autumn-Blooming Herbaceous Plants

Canada goldenrod (*Solidago canadensis*)

Western mountain aster (*Symphyotrichum spathulatum*)

Maximilian sunflower (*Helianthus maximiliani*)

Trees and Shrubs

Golden currant (*Ribes aureum*)

Oceanspray (*Holodiscus discolor*)

Oregon grape (*Mahonia aquifolium*)

Eriogonum ovalifolium

Solidago canadensis

LOW-COST ORNAMENTALS
FOR POLLINATOR GARDENS

It is not necessary to grow only native plants. Including certain nonnatives along with natives can provide excellent pollinator habitat. Look for members of the Mint, Aster, and Rose families, and consider including both perennial and annual plants. The following plants are readily available species that are suited to most parts of the United States.

PERENNIAL PLANTS

Catnip (Nepeta *spp.*)

Coneflower (Echinacea *spp.*)

Lavender (Lavandula *spp.*)

Giant hyssop (Agastache *spp.*)

Oregano (Origanum *spp.*)

Russian sage
(Perovskia atriplicifolia)

ANNUAL PLANTS

Borage (Borago officinalis)

Common sunflower
(Helianthus annuus)
*Avoid "pollenless"
varieties.*

Cosmos (Cosmos bipinnatus)

SHRUBS

Rugosa rose (Rosa rugosa)

Pussy willow (Salix discolor)
*Choose male
(pollen-bearing) plants.*

False indigo
(Amorpha fruticosa)

LAWNS

Lawns represent a great opportunity for pollinator habitat in many urban and suburban areas. Turf grass provides no food value and reduces nesting opportunities for many pollinators.

As a result, even small modifications can make a difference. If having a tidy, mowed lawn is important, consider setting your mower blade higher, to 3 or 4 inches (7–10 cm), and letting the clover or violets grow. If you are open to a little more variety in your lawn, plant low-growing flowering plants such as self-heal, bird's-foot trefoil, clover, and others in the lawn. These flowers are compatible with typical lawn maintenance, and will make a much richer habitat for pollinating insects and a more interesting lawn for you.

△ *Benign neglect will produce a pollinator-friendly flowering lawn.*

△ *Siberian squill (*Scilla siberica*)*

Additional pollinator plants that can be incorporated into unmown areas of lawns include blue-eyed grass (*Sisyrinchium* spp.), wild strawberry (*Fragaria virginiana*), Siberian squill (*Scilla siberica*), and dwarf varieties of grape hyacinth (*Muscari* spp.). The latter two species are planted as small bulbs. All of these plants tolerate mowing and foot traffic, although for grape hyacinth you must wait to mow until the foliage turns yellow in order for it to bloom the following year.

Finally, consider what is really a weed. While you probably don't want to actually plant dandelions or ground ivy (*Glechoma hederacea* or *Pilea nummularifolia*), if you can tolerate them, bees will certainly benefit.

△ *Allowing dandelions to grow can benefit bees.*

Creating an Ecolawn

This approach can be taken even further by establishing an ecolawn. Such lawns may include tough, low-growing grasses that do not require mowing, irrigation, or fertilization, such as buffalograss (*Bouteloua dactyloides*). Ecolawns also often include a diverse mix of low-growing perennials such as short-statured varieties of common yarrow (*Achillea millefolium*), common selfheal (*Prunella vulgaris*), and bird's-foot trefoil (*Lotus corniculatus*). These types of lawns are particularly suited to industrial areas, business parks, and other urban edge areas that require low-maintenance landscaping.

△ *Ecolawns, such as the one occupying the understory of this orchard, not only provide season-long pollen and nectar, but also require less mowing and no fertilizer.*

CASE STUDY

Community Gardens, East Harlem and the Bronx, New York

IN A WARM AND SUNNY SPOT, scientists set out pan traps to catch bees amid lush vegetation, and they observe patches of flowers to catch butterflies. But this is no serene wilderness. From the other side of the fence, the noise of New York's bustling streets intrudes, but on this side is the community garden. Here, local life — vegetable growing, barbecues, band practice, checkers matches — continues around the researchers with its usual vibrancy. Community gardens in East Harlem and the Bronx, tiny squares of greenery hemmed in by buildings and hard surfaces, may not be the typical place to study bees, but Kevin Matteson of Fordham University and colleagues are spearheading a movement to better understand bees and other pollinators in New York City.

Dr. Matteson's research discovered more than 50 species of bees in these community gardens, along with numerous species of butterflies, wasps, and flies. Most of the species found in these small, isolated sites — typically less than 10,000 square feet in size and nearly half a mile from the nearest park — are common and adapted to living in disturbed environments. The cabbage white (*Pieris rapae*) was the most frequently seen butterfly and the common eastern bumble bee (*Bombus impatiens*) the most often encountered bee. One group of bees generally absent was the ground-nesters, particularly the mining bees (family Andrenidae). This is likely because of the lack of undisturbed bare soil; the ground was either covered in concrete or cultivated.

There were surprises as well. A caterpillar munching on rue provided proof that the gardens allow the eastern black swallowtail (*Papilio polyxenes*) to reproduce. And a great spangled fritillary (*Speyeria cybele*) — a rare species even in larger parks of the city — was spotted.

Studies in other cities have also shown that a surprisingly high diversity of bees survive in urban greenspaces. Working with John Ascher of the American Museum of Natural History, Kevin Matteson recorded more than 100 bee species in suburban gardens in New York City's northern fringe. In California, Gordon Frankie counted more than 70 species in private gardens in Albany and Berkeley. Cynthia Fenter and Gretchen LeBuhn found 70 species in urban parks in San Francisco. Fragments of desert scrub habitat in urban areas of Tucson, Arizona, were shown by James Cane, Robert Minckley, and their colleagues to support 62 species of bees. Even green roofs in Chicago have bees: Rebecca Tonietto found 21 species.

Even though not as diverse as most natural areas, when supported with diverse plants and habitats, greenspaces in towns and cities are clearly important for pollinators. ❀

Pollinator Conservation on Farms

RESEARCH CONDUCTED across North America demonstrates that farms adjacent to natural areas have a greater number and diversity of native bees, as well as significantly increased pollination thanks to services from those wild bees. This chapter provides information on how farmers can manage the land around their fields to promote and take advantage of these crop pollinators.

* Native bees make a substantial, but often unrecognized, contribution to crop pollination in the United States.

* Undeveloped areas on and close to farms can serve as long-term refuges for native wild pollinators and strengthen resident honey bee colonies.

* Pollinator habitat can double as conservation buffer areas that help stabilize soil and improve water quality.

Until relatively recently, native bees and feral honey bees could meet all of a farmer's pollination needs for orchards, berry patches, squash and melons, forage seed, sunflowers, and other insect-pollinated crops. Most farms were relatively small and close to areas of natural habitat that harbored adequate numbers of pollinators to accomplish the task that now requires managed colonies of honey bees. Nearby natural areas also served as a ready source of new pollinators that could recolonize farms and provide pollination services if insecticide applications killed resident bees.

Today, however, many agricultural landscapes are much more extensive and lack sufficient habitat to support native pollinators. In spite of this reduction in areas of habitat, the value of the pollination services that native bees provide in the United States is estimated to be worth about $3 billion per year. Research shows that native bees still play an important role in crop pollination across North America, so long as landscapes around farms supply forage and nest sites.

△ *Apple (*Malus*) blossom*

▷ *Squash bees (*Peponapis pruinosa*) foraging in a squash flower*

NATIVE BEES AS CROP POLLINATORS

Many native bee species are much more effective than honey bees at pollinating flowers on a bee-per-bee basis. For example, only 250 female blue orchard bees are required to effectively pollinate an acre of apples, a task that would require one to two honey bee colonies — each containing tens of thousands of workers.

UNLIKE HONEY BEES, bumble bees and several other native bees perform buzz-pollination, in which the bee grabs onto a flower's stamens and vibrates her flight muscles, releasing a burst of pollen from the anther. This behavior is highly beneficial for the cross-pollination of blueberries, cranberries, tomatoes, and peppers, among other plants.

There are many reasons for this increased efficiency. Many native bees, such as mason and bumble bees, are active in colder and wetter conditions than honey bees. In addition, foraging behaviors are more diverse among the native bees than in honey bees alone. The examples of mason bees in orchards and alkali bees in alfalfa illustrate this. In many orchard crops, nectar-foraging honey bees often never contact the anthers, unlike blue orchard bees that forage for both pollen and nectar. Alfalfa flowers are shaped in a way that discourages honey bees from foraging, but the alkali bee can easily forage on these flowers. Some native bees specialize in one type of flower; squash bees (genus *Peponapis*), for example, primarily visit flowers from the Squash family (the cucurbits).

Honey bees also use nectar to pack the pollen into their pollen baskets for transport back to the hive. The nectar moistens the pollen and holds it fast. Many native bees, in contrast, use dense patches of hair to transport dry pollen back to their nests. This dry pollen is more readily dislodged when bees land on other flowers, increasing the chances of plant pollination.

PROTECTING EXISTING HABITAT

Farm areas most likely to support native bees include untilled fallow ground, woodlot edges, stream banks, utility easements, and conservation areas, as well as unused land around farm buildings and service areas. All of these areas can provide both forage or nest sites needed by native bees. Simply leaving these areas alone and protecting them from pesticides and tillage are among the most significant conservation steps you can take.

△ *A native plant hedgerow planted between a field and an equipment parking area takes advantage of marginal ground and maintains undisturbed habitat close to crops needing pollination.*

Preserve untilled areas. Other potential pollinator habitat on the farm includes peripheral areas, such as field edges, fence lines, hedgerows, levees, road edges, and banks of drainage ditches. These areas offer both nesting and foraging sites. This habitat — if not tilled fencerow to fencerow, or if mowed only once a year — can remain relatively stable over time and provide pollinator refuge.

Protect poor-quality land. Some of the best places around farms for native pollinators are the worst places for growing crops. For example, areas with the poorest soils may provide some of the best sites for ground-nesting bees, because these animals often prefer nesting in well-drained sand and silt, and soils low in organic content. The edges and corners of irrigated fields that do not receive enough water to support a crop provide excellent sites for growing various pollinator-friendly plants.

Similarly, the corners of fields that use center-pivot irrigation are often unfarmed. Weed control on these corners can include planting them in nonweedy flowers for pollinators and other wildlife, such as game birds. The edges of tailwater (retention) ponds are additional potential sites for such plants, as are old cemeteries and abandoned homesites.

CREATING NEW HABITAT FOR CROP-POLLINATING BEES

Farmers who want to take a more active role in increasing populations of resident native bees can increase the available foraging habitat to include a range of plants that bloom and provide abundant sources of pollen and nectar throughout spring, summer, and fall. Easy ways to do this include growing appropriate cover crops and letting them bloom or devoting some areas of land to specialized bee pastures.

Bee pastures. A **bee pasture** is land managed for plants that maximize bee reproduction. To be effective a pasture must

△ *Field edges, pond margins, and other noncrop land can be converted into flowering bee pastures.*

provide an abundant bloom throughout the nesting period, especially in the early stages of bee emergence. Such plantings typically consist of high-density wildflower meadows with a diversity of plant species. Ideally, bee pastures are comprised of locally appropriate native plants, but mixtures of native and nonnative species — especially nonnative species with prolific blooms, such as yellow sweet clover, canola, purple vetch, red clover, buckwheat, and alfalfa — are acceptable when cost or seed availability limits native plant choices.

Cover crops. The ground beneath the rows of berry crops and vineyards, as well as fallowed fields, can easily be sown in a ground cover that provides nectar and pollen. These cover crops can also provide several other benefits, such as improving erosion control and soil permeability, fixing nitrogen, discouraging weeds, and harboring beneficial insects.

Consult with local seed suppliers or the local Cooperative Extension Service to determine cover crops best suited to a particular area. The best plants for a cover crop are short with abundant flowers. Some excellent choices for bees include varieties of clover, trefoil, vetch, scorpion weed (*Phacelia*), and buckwheat. Be sure to wait until the flowers have bloomed before tilling or mowing these cover crops.

12 Pollinator Conservation in Natural Areas

* Many types of wildlife benefit from intact pollinator communities.

* Wild lands support native pollinators that can provide crop pollination on adjacent farmland.

* Specialist pollinators depend on specific host plants and nest conditions typically found only in natural areas.

SUNNY, open wild areas are often a mosaic of distinct plant communities — such as oak savannah, scrublands, and prairies — in which pollinators thrive. Additionally, transition areas, such as open woodland or forest edges, often contain mature trees, snags, and fallen logs that provide nest sites, as well as shrubs and other small plants that offer foraging opportunities for pollinators.

Pollinators Are Essential For Healthy Ecosystems

Natural areas are not just important for pollinators — pollinators are also vital to the functioning of these ecosystems. Pollinators are important in wildlife food webs both as an essential step in the production of seeds, nuts, fruit, and berries, and as direct prey. Bears, rodents, small mammals, birds, and many terrestrial invertebrates all depend directly or indirectly on pollinators for much of their diet.

Although in almost all cases pollination is carried out by adult insects, both adults and larvae play other essential roles in ecosystems. All of these insects feed songbirds, game birds, and fish, decompose detritus, and act as pest control agents. A strong case can be made that pollinators are a keystone group in nearly all terrestrial ecosystems, necessary not only for plant reproduction but also for forming the basis of an energy-rich food web. Pollinator conservation, in short, provides a framework for enhancing and managing natural areas with benefits for a wide variety of other invertebrates that subsequently are invaluable for wildlife.

▽ *Pollinators are a keystone group of animals in natural areas where they are necessary for the reproduction of many trees, shrubs, and wildflowers.*

CASE STUDY
Oak Savanna Restoration, Oregon

A FEW YEARS AGO, walking in a straight line across the hilltop on Jefferson Farm would have been impossible, at least without a brush cutter. Any views into Oregon's Willamette Valley would have been blocked by decades of scrubby growth. Now, mature oaks stand scattered across acres of savanna, bird song fills the air, and butterflies and bees move from bloom to bloom across the sun-warmed grassland. This change did not happen overnight: it is the result of much hard work by the site's owners, Mark and Jolly Krautmann, and their staff, principally Lynda Boyer.

Once widespread across the Willamette Valley, only about 1 percent of the Valley's oak savanna remains. Historically, the savanna was a flower-rich grassland with only five to ten oak trees per acre. In the case of Jefferson Farm, the land was heavily degraded by decades of grazing, and much of the savanna was smothered by dense oak thickets and brush. With the aim of returning the site to its historic condition, the Krautmanns decided to control the nonnative and invasive weeds with herbicides and clear the thickets by cutting and brush mowing. Large conifer trees were limbed and topped, creating snags in which birds and insects can nest.

Beneath the oaks, grassland supporting more than 50 species of native wildflowers and grasses is being restored. A variety of seeding methods have been used on different parts of the site, including a seed drill, a pull-behind spinner spreader, hydroseeding in steep slope areas, and hand seeding. Some patches were left unseeded, including sites known to harbor ground-nesting bees.

The restoration project will require ongoing monitoring and management. Fire will be the main tool for managing the grasslands. To minimize disruption to local pollinators, the site has been divided into a series of smaller units that will be burned on a rotation of 3 to 5 years.

Projects like this do not come cheap. The Krautmanns committed significant resources to it and received financial support from federal and state wildlife agencies. They also engaged a variety of organizations, including the U.S. Fish and Wildlife Service, Willamette University researchers, the Xerces Society, public schools, and community groups.

Natural areas like Jefferson Farm are gems in the landscape, worthy of protection and management in their own right, but they also serve as an important source of native bees for the surrounding area. These pollinators can visit adjacent crops or colonize new habitat created for pollinators in nearby intensively farmed landscapes. ❀

Urban Greenspaces, Parks, and Golf Courses

13

I N AN INCREASINGLY urbanized nation, the greenspaces scattered through urban and suburban areas — varying from river corridors and urban trails to golf courses and carefully manicured city parks — contribute to the vitality of our communities. In addition to supplying a welcome break from the hard surfaces of neighborhoods and business districts, these areas provide a wealth of opportunities for recreation and relaxation, aid community development, and help maintain a clean, healthy environment.

✿ Pollinator conservation is well suited to the small-scale patchy landscapes of urban and suburban greenspaces.

✿ Providing habitat for pollinators helps gardeners, as well as local farms and natural areas.

✿ Habitat patches enhance environmental education programs and visits to parks.

✿ Pollinator conservation efforts may help parks or golf courses achieve "green" certification.

If you do use pesticides, carefully follow label guidelines and take steps to minimize impact on pollinators, such as:

* Use the least-toxic product for the target pest.
* Apply pesticides only on windless or slightly breezy days to avoid drift into habitat areas or into neighborhoods.
* Create wide buffer zones to protect adjacent habitat and communities.
* Apply insecticides at night when pollinators are not active.

Mowing Around Pollinators

Mowing is another management activity that can harm pollinators. One simple way to increase habitat is to alter how, when, or where mowing is done. By reducing the frequency of mowing, weedy flowers such as dandelions and clover will be able to bloom periodically. By raising the height of the mower, these and other low-growing flowers have a better chance of reaching blooming height. Reconsider whether some areas can be left unmown, such as field corners or cart-path margins. For wilder areas managed as pollinator or wildlife habitat, set up a staggered mowing schedule, cutting only a third of the pollinator areas or out-of-play areas each year. Fire is an alternative to mowing that may be appropriate for some sites.

Land Managers as Leaders

Care for the environment has been at the core of the park managers' and golf course superintendents' professions for many years. Increasingly, local communities, park users, and golfers are looking to managers and superintendents to take a lead in wildlife conservation and environmental stewardship. Maintaining pollinator populations is one of the most valuable ways in which these areas can contribute to a healthy environment.

Special Considerations for Other Landscapes

IN ADDITION TO THE LANDSCAPES discussed in the preceding chapters, many other places can support the pollinator populations that are central to a healthy ecosystem. These areas include business and school campuses and marginally managed greenspaces near roads and industrial areas. Such human-dominated settings are often the last places considered for wildlife conservation efforts. Yet the incorporation of pollinator habitat into these places can have direct aesthetic and economic benefits.

* Pollinator conservation can be adapted to any available space.

* Pollinator conservation provides opportunities to build community and raise environmental awareness.

* Efforts to provide for pollinators in a variety of greenspaces will also attract and support other wildlife.

* Managing industrial and urban greenspaces for pollinators can lower land management costs and energy use.

CASE STUDY
Bonneville Power Administration, Oregon

POLLINATOR CONSERVATION is a relatively straightforward process in many landscapes where the goal is simply to recognize and protect existing habitat. This "do no harm" approach becomes much more challenging, however, when the landscape has to accommodate heavy equipment traffic and disruptive management activities. That is the situation faced by the Bonneville Power Administration along its entire transmission network in Oregon. Public safety and service reliability require BPA to remove tall trees below the power lines, and to periodically upgrade or repair the tower structures.

Balancing these issues has become even more important along the Santiam-Toledo Transmission Line Corridor in western Oregon, which supports an important population of the imperiled Taylor's checkerspot butterfly (*Euphydryas editha taylori*). The corridor provides a landscape similar to the now very rare western Oregon prairie ecosystem that was once abundant in the low hills flanking the Willamette Valley.

Since European settlement, more than 99 percent of this prairie habitat has been lost, either to agriculture or to land development. Prairies of this type also have disappeared due to a change in fire management policy. Suppressing wildfires has allowed open prairie areas to become overgrown with trees and invasive plant species such as Himalayan blackberry, thistle, false brome, and Scotch broom.

Although BPA's vegetation management practices help prevent the encroachment of these invasive plants, its maintenance activities presented a potential threat to the butterfly's nectar and host plants. To find an appropriate solution to these varying demands on the transmission line, BPA worked with the Xerces Society and the U.S. Fish and Wildlife Service to develop a right-of-way management plan that balanced corridor maintenance with butterfly protection.

The plan called for educating BPA staff about the presence and needs of the butterfly, and documenting especially important sections of the corridor on maintenance maps. Using this information, BPA schedules maintenance activities along those critical sections so that they occur during the periods of the year when the larval and adult stages of the butterfly are least active.

BPA also has modified vegetation management efforts to limit overspray of herbicides onto critical habitat areas, and to target only noxious weeds. Where areas must be revegetated, BPA uses native, noninvasive plants of particular importance to the butterfly, such as wild strawberry (*Fragaria virginiana*), mariposa lily (*Calochortus* spp.), and rosy plectritis (*Plectritis congesta*), as well as stonecrop (*Sedum* spp.), scorpion weed (*Phacelia integrifolia* spp.), and pussypaws (*Cistanthe* spp.).

Preservation of existing prairie habitat and restoration of degraded areas is the primary conservation need for the Taylor's checkerspot in western Oregon. BPA's efforts alone will not restore those large areas of historic habitat, but they do serve as an example of how one organization can reduce its ecological footprint to help the survival of a highly imperiled species. ❧

Grassroots Action 15

THIS BOOK EMPHASIZES what you as an individual can do to protect and strengthen pollinator populations by guiding you through the steps necessary to manage or enhance pollinator habitat in your backyard, farm, school campus, or natural area. However, to protect pollinators for the long term it's important to consider these species and their ecological role in a larger context, and to enact policies that protect them on a local, national, and international scale.

❁ Talk to your friends, relatives, and neighbors about the role of pollinators in human food production and ecosystem health.

❁ Be alert for policy issues that affect pollinators. Let your government representatives know how you feel.

❁ Join local efforts to protect and restore natural areas and monitor pollinators.

WHAT THE PUBLIC CAN DO

Citizens have many ways to affect policy. Contacting Congress to urge support of appropriating money for research into the problems facing honey bees, native bees, and pollinator habitat is one option.

HIGHWAY HABITAT

Every six years, the U.S. Congress passes a transportation bill that establishes funding and infrastructure priorities for road and highway construction and maintenance. These public rights-of-way are superb sites for pollinator habitat if planted with diverse wildflowers and low-growing shrubs. These areas sown in wildflowers require less maintenance (mowing), create markets for native seed producers, and create corridors of habitat across the United States. Not only that, but they are beautiful.

WORKING WITH YOUR STATE

It is also important to educate yourself about how you can influence or educate state agencies. These include state departments of transportation, agriculture, and natural resources. Contact these state policy makers and land-use managers and encourage them to include diverse plants and nest sites for pollinators in the lands they manage.

Consider contacting your state department responsible for pesticide regulation as well (sometimes this is under the state department of agriculture) to ask what they are doing to help educate licensed pesticide applicators about reducing their impact on pollinators. Ask the regulators to include or develop special trainings on protecting pollinators from pesticides and to strictly enforce label requirements about protecting bees.

GOING LOCAL

You can also work with local conservation groups to protect and expand open space areas, wildlife refuges, and wilderness areas. Find out who manages the plants and animals in your neighborhood greenspaces or nearby open areas, and encourage them to include pollinators in their management plans. Controlled burns, for example, though an important tool for maintaining open prairie, grassland, and meadow habitat, must be timed and managed with pollinators in mind lest they decimate their populations. Work with local park managers to minimize the use of pesticides. Contact your local mosquito-abatement district to learn what it is doing to minimize the impact of its spraying on pollinators.

You also can help educate your community about the importance of pollinators. You can teach teachers, students, friends, neighbors, colleagues — and anyone else who will listen — about pollination. Help your elected officials understand the crucial role that pollinators play in urban yards and gardens. Watch for development proposals that might affect an adjacent natural area and then talk to city or regional planners about protecting it. Contact the agencies responsible for highway margins and advocate that maintenance crews cut only bushes and higher vegetation, and only where public safety demands it. Suggest that native plants and seeds be used in these public areas.

These might seem like small steps to address a big problem, but together the actions of individual citizens in their local parks, neighborhoods, and backyards can have a significant impact. Everything you do will make a difference.

▽ *Organizing walks or other events to look at good habitat or talk about potential improvements can be an excellent way to engage local communities.*

CASE STUDY
The Great Sunflower Project, San Francisco, California

TWICE A MONTH, on a sunny day, take a lawn chair into your garden and sit in it for 15 minutes watching the sunflowers. This might sound like a relaxation exercise recommended by your doctor, but in fact, watching sunflowers has become one way in which tens of thousands of people across North America have been able to contribute to the understanding of what's happening to bees across the continent.

The Great Sunflower Project, a citizen-science project run by Dr. Gretchen LeBuhn of San Francisco State University, has unlocked the secrets of backyards in all 50 U.S. states and all provinces and territories in Canada. The project is based upon a simple methodology. People grow sunflowers, gather data on the bees that visit them once they bloom, and then submit the data to the project via its website.

The sunflowers are all of the same variety, annual 'Lemon Queen,' its seeds mailed to them by the project managers. Using the same variety means data from different gardens are comparable. This has provided Dr. LeBuhn with a massive amount of data, giving insights into which areas have lots of bees or only a few (and thus which areas are receiving less pollination by bees), as well as which factors influence the presence or absence of bees.

Credit for the project's success most certainly goes to Dr. LeBuhn, and other organizations have contributed to this project as well. The Xerces Society partnered to develop an identification guide for bees, and the American Museum of Natural History and the San Francisco Garden Resource Organization collaborated to plant flowers in San Francisco's community gardens. The Great Sunflower Project also joined with Cornell University on The Birds & the Bees Challenge, which focuses on getting young people to explore nature in their own neighborhoods.

From its earliest days in 2008, The Great Sunflower Project has attracted huge interest. Fifteen thousand people registered in the first week, and by the second year, more than 50,000 people signed up to receive seed packets from the project. Thus, in addition to the scientific data, the project has created a virtual community of like-minded, concerned citizens. They can share stories and photos via the project's website, discuss sightings, and glean advice on what to do to support bees. Generating this kind of enthusiasm and action has been an unintended benefit from relaxing in the sun and observing what is — or is not — all around. ✿

POLLINATOR SCRAPBOOK

PART 3

Bees of North America

This section will help you to identify some of the more abundant or important bees found in parks and gardens, on farms, and in natural areas, as well as to better understand their natural history and conservation needs. Novice birders take some time to learn birds (and they are a comparatively easy group of animals!) so don't despair if "bee-ing" takes some time to get used to as well.

Globally, bees are grouped into seven families, and although six of those are found in North America, only five are common. Bees usually can be distinguished from other insects by their two pairs of wings and the presence of branched hairs, plus pollen-collecting structures on females. Birders talk about simple characteristics, such as a bird's stance or bill size or the way it holds its head, that enable them to identify a species instantly. Many bees also have a similarly distinctive aspect. Bumble bees are round and hairy. Long-horned bees have long antennae. Metallic sweat bees are a bright metallic green. While some bees will always be tricky to recognize, with practice others will quickly become obvious.

PART 3 IS DIVIDED into three sections. First is an overview of the diversity and taxonomy of bees. The second section shows how to separate bees from flies, wasps, and other insects: is it a bee you're looking at? The last and largest section, the identification guide, consists of profiles of bees with color photos. Each profile offers a description of a bee genus, accompanied by notes on its distribution, nesting and foraging habits, and any conservation concerns. Scientific language is used in some places for precision; the glossary on page 345 in the appendix explains all the technical words.

This book will not help you to identify bees to species level (except for the European honey bee). Identifying bees in the field is difficult. To separate species and even some genera, in some cases, you need to see what shape the tips of hairs are, which is not something that can be done easily without a microscope. Fortunately, for all but the most detailed of scientific studies, knowing the species is not necessary. Being able to assign a genus, even a family, can tell you a lot about the bees' nesting needs and foraging habits, information that has practical uses for gardeners, land managers, and other people wanting to plan habitat projects.

BEES ARE EVERYWHERE

Taxonomists estimate that there are 20,000 or more species of bees in the world, with 4,000 species in North America. The diversity of bees on this continent includes bumble bees, carpenter bees, mason bees, sunflower bees, leafcutter bees, squash bees, alkali bees, sweat bees, and long-horned bees, to name just a few, as well as the European honey bee and a handful of other nonnative species.

North America's bees are incredibly diverse in color, shape, size, and natural history. The smallest are ground-nesting *Perdita*, which may be no longer than 1/12 inch (2 mm), hairless, and brown or black with yellow markings. The largest are carpenter bees and bumble bees, which may be an inch (25 mm) or more long and are round and hairy. Some bees live in colonies, others on their own; some dig nesting tunnels in the ground and others occupy hollow plant stems. Bees may be black or brown or sparkling emerald green. They may have markings in red, yellow, gray, white, or flame orange.

Bees live anywhere there are flowers offering nectar and pollen. They tend to be more diverse and abundant in warm-temperate and Mediterranean climates, and less so in the humid tropics and colder regions (higher altitudes or more northern latitudes). At the continental scale, there are far more species of bees in the drier areas west of the Rockies than there are in the moister, temperate areas to the east. California is the state with the highest diversity of bees: almost 2,000 species. Indeed, the place with the highest known diversity of bee species in the world — more than 400 species — is Pinnacles National Monument in California.

This is an exceptional number of species in a protected area, but even small or apparently less noteworthy locations support significant bee diversity. For instance, two studies found more than 200 species around Carlinville, Illinois, a landscape of mixed farming and riparian woodland. Bees also survive in developed areas: more than 100 species were found in gardens in the northern suburbs of New York, and more than 50 in community gardens in Harlem. Wherever you look, you can find bees.

THE NAME GAME

Look in a field guide to butterflies, grasshoppers, or dragonflies, and every species will have a common name. This is not the case for bees. Most lack a common name, and some names are applied to dozens of species — and other bees have multiple common names! Obviously, this can lead to confusion. It also means there is a greater reliance on scientific names with bees than other groups of commonly studied animals.

Each bee has a two-part name comprising a genus name (describing a group of closely related bees) and a species name (describing one particular type of bee). This genus-species combination forms the scientific name, called a **binomial**, which is unique for each species. It is not used for any other living organism. The genus and species names are at the bottom of a hierarchical naming system that starts with the broadest category of kingdom and becomes increasingly restrictive down through phylum, class, order, and family to genus and species (and eventually subspecies in some cases). There are also further subdivisions within these categories, such as superfamily, tribe, and subgenus.

Not only is the naming system standard, but the way the names are written also follows certain rules. All names except the species and subspecies begin with a capital letter, and all but genus, species, and subspecies are written in normal type.

△ *Mining bees (genus* Andrena*) often carry pollen high on the rear legs and the rear of the thorax.*

and *Perdita*, which include more than 400 and 600 species, respectively. Andrenidae are abundant in the spring, especially in dry temperate regions, and are among the first bees to emerge in late winter.

All andrenids excavate nests in the ground, giving them their common name of mining bees. They generally nest in flat or gently sloping sites and may form aggregations with tens of thousands of bees nesting in a small area.

The huge majority of species in this family are solitary. A small number of species are communal, with several females sharing a nest entrance but each excavating and provisioning their own brood cells. Bees in the genus *Perdita* do not line their brood cells. All the other species do, though the lining soaks into the soil (unlike the Colletidae). This family is unusual because it does not include any cleptoparasitic species. (**Cleptoparasites** do not make their own nest, but lay eggs in the nests of other species.)

The three andrenid genera in this guide are *Andrena*, *Perdita*, and *Protoxaea*.

Halictidae (Sweat Bees)

Halictidae are probably the most frequently encountered bees during summer. Although this family includes most of the brightly colored metallic bees, the majority of halictids are drably colored and small. Since they don't fit the widely accepted image of a bee, they are generally overlooked.

The nesting behaviors of the Halictidae vary greatly. Most excavate nests in the ground, though some create them in rotting wood. The genus *Lasioglossum* includes the whole gamut of social behaviors from **solitary** (each female constructs her own nest) to **semisocial** (nests are constructed by a group of bees in which one is the egg-laying queen). Some *Agaposte-mon* are **communal**; a dozen or more females may share a nest entrance, but underground each bee creates her own brood

△ *Sweat bees typically appear dark and drab, but they are among the most abundant bees during summer.*

cells. Species in the genus *Sphecodes* are all cleptoparasites, mainly in the nests of other halictids, but occasionally in nests of *Andrena* or *Perdita* bees.

The halictid genera in this guide are *Nomia, Dieunomia, Agapostemon, Halictus, Lasioglossum, Sphecodes,* and *Augochlora.*

Melittidae (Oil-Collecting Bees)

The Melittidae is a small family of bees found only in temperate regions of the Northern Hemisphere. In North America, there are only three genera: *Hesperapis, Macropis,* and *Melitta.* Of these, *Hesperapis* is the most common, but found only in the Southwest; the other two are rare. They share few characteristics to aid in identifying them. The most obvious feature is the pollen brush (scopa) on the hind legs; it extends farther down the leg than in other bees.

All melittids are solitary bees and nest in the ground, and they typically leave the brood cells unlined. *Macropis* is the exception; it collects oils from yellow loosestrife (*Lysimachia*) with which it lines the cells. It also mixes the oil with *Lysimachia* pollen to provision the cells.

Since these bees are generally rare, this guide does not include any melittid genera.

important to identify different bee families and species. There are two large, compound eyes and three small simple eyes (**ocelli**) on the top of the head. The antennae are segmented, cylindrical, and bent.

THE THORAX

Wings and legs are all attached to the **thorax**, the central major body region. One of the defining features of bees is that they

have four wings (technically, two pairs). The front edge of the hind wing has a structure called a **hamuli**, which looks rather like a single row of Velcro hooks, connecting it to the forewing. Thus, unlike a dragonfly, for example, on which each of the four wings moves independently, the two wings on each side of a bee flap together and at a glance appear to be a single wing. At rest, bees may lay their wings across their backs

△ The thorax is the region to which most appendages are attached: two pairs of wings and three pairs of legs.

▽▷ Pollen is transported either under the abdomen (R) or on the hind legs (L), though on most bees it is carried as a dry powder.

but do not fold them. When viewing a bee under a hand lens or microscope, the shape of veins and cells is important for species identification, but this is not much use when watching a bee on a flower!

Of a bee's three pairs of legs, the hind pair is the one that is most useful for identification — and easiest to see. As noted above, Megachilidae is the only family in which bees transport

pollen in hairs on the abdomen; bees in all the other families have structures on their hind legs for carrying pollen. These structures are really easy to spot when the bee is carrying pollen, but when empty they are either a patch of long hair (making some legs look rather like a bottlebrush) or a broad area of bare, shiny leg surrounded by incurving, bristly hairs. (The exception is the Colletidae genus *Hylaeus*, which carries pollen internally; other colletids carry it on their legs.)

The bristly body hair is branched, which helps pollen stay on the bee, as does a second trait, an electrostatic charge. Research has shown that the accumulated charges on bees are sufficient for pollen to be deposited on oppositely charged flowers by electrostatic discharge.

THE ABDOMEN

The third major region, the abdomen, is connected to the thorax by a very narrow waist and ends (in females) with the stinger. Some bees have colored markings or stripes on their integuments; others have bands of colored hairs. The patterns or positions of these can be very useful in identifying bees. Carder bees, for example, have very distinctive color patterns that allow confident identification at a glance.

The terminology used to describe the abdomen in some books can be confusing. Technically speaking, the rear section of the thorax (**propedeum**) is actually the first segment of the

---

MALE OR FEMALE?

Once you begin spending time observing bees, you'll notice many different types. Most of the information included in this guide relates to females. Male bees frequently have different markings, and they never have baskets (**corbicula**) or brushes (**scopae**) for carrying pollen. Because they produce eggs, females are usually larger and more robust than males of the same species.

The chart below highlights the principal differences between the sexes.

	FEMALE	MALE
Pollen transport structure*	yes	no
Antenna segments	twelve	thirteen
Abdomen segments	six	seven
Stinger	yes	no

*Note that female cleptoparasitic bees lack pollen-collecting structures, also those in the genus *Hylaeus*.

SUBFAMILY: Andreninae

Andrena: Mining Bees

(an-DREE-nuh)

0.3 TO 0.7 INCHES
(7–18 MM) LONG

ANDRENA ARE FREQUENTLY ENCOUNTERED by gardeners because of their habit of nesting in lawns. This is a large and diverse genus with nearly 1,500 species worldwide, and about 400 in North America. Though they are found in all types of habitat, *Andrena* are much more common in temperate than tropical regions.

IDENTIFICATION: Small to moderate-sized bees, 0.3 to 0.7 inches (7–18 mm) long. Most North American species are black, dull metallic blue, or green and moderately hairy, with bands of pale hair on their abdomens. Females have large, velvety facial depressions (foveae) that look like eyebrows and large pollen-collecting hairs (scopae) on the upper part of their hind legs, seemingly in their "armpits." Despite a variety of striking colorations, *Andrena* species are difficult to tell apart.

SIMILAR INSECTS: *Colletes* and *Halictus* can be a similar size and color, but the lack of stripes on the abdomen of *Andrena* makes it straightforward to separate them.

FORAGING: In the western United States, *Andrena* are among the most common bees that forage in the spring, with some species emerging as early as February. The genus contains both species that visit a wide variety of flowers and specialists.

NESTS: All *Andrena* nest in the ground, typically in sandy soil and often near or under shrubs. The nest entrance is usually marked by a small mound (tumulus) of soil. The female lines her brood cells with a material she secretes from a specialized gland. Unlike *Colletes*, the secretion of *Andrena* soaks into the soil. It dries and is rubbed by the female using a flat area near the tip of her abdomen (the pygidial plate) to form a highly polished and stable cell wall.

Many *Andrena* species nest in large aggregations of tens of thousands of bees. Even within such a large gathering, the individual bees are solitary, though nests can be in densities of 30 entrances per square foot. A few species nest communally, where two or more females share a nest entrance but build and provision their own brood cells.

CONSERVATION CONCERNS: Several *Andrena* species are considered vulnerable or imperiled. *A. aculeata* and *A. winnemuccana* are on the Xerces Society's Red List (see page 13), though little is known about their biology. Both are limited to the Columbia River basin. Only one specimen of *A. winnemuccana* has been collected, and both species may be threatened by habitat loss. Several *Andrena* species from California and Nevada are listed as critically imperiled or imperiled on the NatureServe Explorer database (see page 352 in appendix).

DID YOU KNOW? *Andrena* regularly nest in lawns, especially in spring, but pose no risk to people because their weak stingers cannot penetrate human skin.

SUBFAMILY: Panurginae

Perdita: Miner Bees

(per-DIH-tuh)

0.1 TO 0.4 INCHES
(2–10 MM) LONG

A DIVERSE GENUS of more than 600 species, *Perdita* is restricted to the Americas from Guatemala to southern Canada. The greatest abundance and diversity of *Perdita* species are in the deserts of the southwestern United States and northern Mexico. These include the tiniest bees found in North America.

IDENTIFICATION: Tiny to small bees, 0.1 to 0.4 inches (2–10 mm) long. Most are black or metallic green or blue, frequently with yellow or white markings. In some species, these markings expand until they appear to cover the entire body.

SIMILAR INSECTS: Easily mistaken for tiny dark wasps and dark flies.

FORAGING: All are specialist foragers. Each species relies on a small number of flower species for nectar and pollen. Collectively, they utilize a wide range of plants including willows, lotus, asters, sunflowers, and members of the Spurge and Phlox families. Females typically carry dry pollen on their hind legs, but those in at least one subgenus, *Pygoperdita*, carry pollen moistened with nectar.

NESTS: As with other bees in the family Andrenidae, all *Perdita* nest in the ground. Most are solitary but some nest communally with many females sharing the same nest. *Perdita* is unique among ground-nesting bees because it does not line its brood cells. To protect the food from fungal attack, the females cover the stored pollen, instead of the brood-cell walls, with a glandular secretion. The newly hatched larva chews through this covering to reach the food. As the larva grows and the

food mass (and its protective layer) shrinks, the larva will roll onto its back and hold the food mass on its belly. Before rolling over, the larva develops fleshy growths along its back to raise it up off the cell floor.

CONSERVATION CONCERNS: Several *Perdita* species are listed as vulnerable on the Xerces Society's Red List (see page 13). Several others are listed, but not yet ranked. Most of these have been collected only once or twice, and all are endemic to the Columbia River basin. Additional species from California and Nevada are listed as critically imperiled or imperiled on the NatureServe Explorer database (see page 352 in appendix).

DID YOU KNOW? This genus includes the smallest bee in North America, *Perdita minima*, found in the deserts of the southwestern United States. It is only 1/12 inch (2 mm) long.

SUBFAMILY: Megachilinae

Hoplitis: Mason Bees

(hop-LIE-tiss)

0.2 TO 0.6 INCHES
(5–15 MM) LONG

HOPLITIS ARE CALLED MASON BEES because they construct walls to separate the brood cells in their nests. It is a genus with about 60 species in North America. They are more common in the western states.

IDENTIFICATION: Slender to robust bees, *Hoplitis* range from small to moderate-sized, 0.2 to 0.6 inches (5–15 mm) long. They are often black, but some species have a red abdomen with bands of pale hair; a few species are metallic. *Hoplitis* females carry dry pollen in patches of hair (scopae) on the underside of the abdomen.

SIMILAR INSECTS: May be confused with *Osmia*, but on the whole most *Osmia* are metallic, while many *Hoplitis* are not.

FORAGING: *Hoplitis* fly primarily in the spring. Most species visit a wide variety of flowers, with many favoring plants in the Pea, Mint, and Figwort families; some are specialists.

NESTS: *Hoplitis* are solitary nesting bees. They are opportunistic and use a variety of sites including pithy stems, holes in wood (both regular and irregular shaped), nests of other insects in the soil, or above-ground nests of mud-dauber wasps. The nest holes are divided into brood cells with walls made of a diversity of materials, including chewed leaves, pebbles, sand, clay, and bits of wood collected from outside the nest.

CONSERVATION CONCERNS: *Hoplitis orthognatha* and *H. producta subgracilis* have very limited ranges within the Columbia River basin. *H. orthognatha* has been recorded in only three locations. Both are listed as vulnerable on the Xerces Society's Red List (see page 13). Some *Hoplitis* species and other bee pollinators of an endemic vetch in California may be at risk due to sheep grazing; one study found that sheep destroyed nests, removed bee food sources, and even trampled the bees themselves.

DID YOU KNOW? In Europe, there are several species of *Hoplitis* that nest in old snail shells. One of these even uses sheep or rabbit dung to line the brood cells!

SUBFAMILY: Megachilinae

Chelostoma: Mason Bees
(chel-AH-stoh-mah)

0.1 TO 0.4 INCHES
(3–10 MM) LONG

GLOBALLY, *CHELOSTOMA* IS FOUND in Asia and Europe, as well as North America, where it is limited to 11 species (9 native and 2 nonnative). Most North American *Chelostoma* species are found in the west of the continent. Just one native and two nonnative species are found east of the Rocky Mountains.

IDENTIFICATION: Very slender, small black bees from 0.1 to 0.4 inches (3–10 mm) long, with only one species attaining a larger size of 0.5 inches (13 mm). *Chelostoma* females carry pollen in scopa on the underside of the abdomen.

SIMILAR INSECTS: *Chelostoma* may be confused with *Hoplitis* and *Osmia*. The bees in these other two genera are typically more robust and rounder, and are sometimes metallic blue, green, or red in color.

FORAGING: Most (perhaps all) *Chelostoma* bees are floral specialists, foraging for pollen from only specific plants. The eight species in western North America are believed to be attracted to plants in the Hydrophyllaceae family, especially the genera *Phacelia* and *Eriodictyon*. The one eastern native species (*C. philadelphi*) is a specialist on mock orange (*Philadelphus*).

NESTS: *Chelostoma* are tunnel-nesting bees like other mason bees such as *Osmia* and *Hoplitis*. They select hollow stems and abandoned beetle borer holes, in which they create and provision a linear series of nest chambers. They construct partitions between chambers with extremely hard cementlike mixtures of fine pebbles and plant resins. Sometimes they line the interior surfaces of nests with a water-resistant varnishlike glandular secretion.

CONSERVATION CONCERNS: One nonnative species, *Chelostoma campanularum*, is a specialist pollinator of common bellflower (*Campanula rapunculoides*), a nonnative invasive plant that has escaped from gardens in many areas of the eastern United States. The presence of this bee may help increase the spread of the plant.

DID YOU KNOW? Several *Chelostoma* species collect flower nectar to use as a glue to hold together the sand grains and small pebbles that form the partitions between nest cells.

SUBFAMILY: Megachilinae

Megachile: Leafcutter Bees

(meg-uh-KILE-ee)

0.4 TO 0.8 INCHES
(10-20 MM) LONG

MEGACHILE CUT PIECES OF LEAVES or petals with which to construct their brood cells, hence their common name of leafcutter bees. The genus is large, containing nearly 1,100 species worldwide and about 140 in North America.

IDENTIFICATION: Moderate to large bees, 0.4 to 0.8 inches (10–20 mm) long. *Megachile* are smoky-colored, stout-bodied, and have flattened abdomens with pale hair bands. Females often have a broad head to accommodate large mandibles for cutting leaves. The broad head and stout body, combined with an often-upturned abdomen, can give them a pugnacious look. Females carry dry pollen on a patch of hair (scopa) on the underside of their abdomens.

SIMILAR INSECTS: Nonmetallic *Osmia* species, such as the nonnative *O. cornifrons,* closely resemble *Megachile* bees. To tell them apart, look for a generally more upturned abdomen on megachilids.

FORAGING: *Megachile* includes species that visit a wide variety of flowers as well as species that specialize on certain plants such as those in the Aster and Pea families. The alfalfa leafcutter bee (*M. rotundata,* an introduced species from Eurasia) is used commercially to pollinate alfalfa.

NESTS: *Megachile* are all solitary. Almost all North American species nest in preexisting cavities, either natural or artificial: abandoned beetle tunnels, hollow plant stems, gaps behind loose bark or stones, nest blocks, hose pipes, and any other suitable hole. A few species dig nests in the ground, often in loose sandy soil. This is more common for species in the desert Southwest but *M. addenda* builds nests in the sandy banks adjacent to cranberry bogs along the Atlantic coast.

These bees wrap brood cells in leaf or petal pieces and prefer leaves that are smooth on one side, to face into the cell. The female cuts pieces of particular sizes or shapes to use in different parts of the cell. For example, the cell's sides are made with oval pieces, and it is closed with a number of circular pieces. Gardeners often believe that *Megachile* are harmful to plants, even though their leaf-cutting and stem-nesting activities cause only cosmetic damage.

CONSERVATION CONCERNS: The nonnative *Megachile apicalis* is believed to usurp the nests of native leafcutter bees in parts of California and the Southwest. It is also a specialist pollinator of yellow star-thistle (*Centaurea solstitialis*) and may contribute to the spread of this invasive weed.

DID YOU KNOW? It only takes two or three seconds for a female *Megachile* to cut a piece of leaf. Just before she finishes cutting the leaf, the female starts to beat her wings so that she is already flying by the time the leaf fragment is severed.

SUBFAMILY: Xylocopinae

Xylocopa: Large Carpenter Bees
(zy-low-COPE-uh)

**0.5 TO 1.25 INCHES
(13–30 MM) LONG**

XYLOCOPA **ARE LARGE BEES,** frequently confused with bumble bees. Many people feel intimidated by the size of *Xylocopa*, but they are gentle creatures. Globally, there are several hundred species, but they are largely tropical bees. Only a handful of species are found in North America.

IDENTIFICATION: Robust, often with a hairy thorax but only sparse hairs on abdomen. Moderate to large, ½ to 1¼ inch (13–30 mm) long. Most are black, though they may have a metallic sheen of blue or green with blackish wings. Males of some species are sometimes golden brown or brick-colored. Females carry dry pollen on scopae on their legs. Male *Xylocopa* are relatively territorial and may buzz around humans. Like all male bees, they are unable to sting.

SIMILAR INSECTS: Large carpenter bees are often confused with bumble bees (*Bombus*) because they are a similar size. However, because *Xylocopa* are much less hairy, they are shiny in comparison, especially on the abdomen.

FORAGING: Large body size limits *Xylocopa* to visiting large or open-faced flowers; the flower needs to be strong enough to support the weight of the bee! However, some species put their strong jaws to good use and chew holes into the sides of flowers (such as blueberries) to rob nectar. *Xylocopa* pollinate many crops, including passion fruit, blackberry, and pepper.

NESTS: Most species of *Xylocopa* make solitary nests. In the few semisocial species, generations overlap, and mothers and daughters share a nest. *Xylocopa* are among the few bees that can excavate their own nests in wood, using powerful jaws to chew into soft wood and plant stems. Those species that nest in stems construct a linear series of brood cells divided by walls of compacted sawdust. (An example is the California carpenter bee, *X. californica*, which uses yucca and agave stems.) Those that chew into soft wood may excavate brood cells off a main tunnel.

Xylocopa are well-known for chewing nest cavities in the structural timber of buildings, though a good coat of paint is all it takes to deter them. *Xylocopa* species that live in desert habitats line their nests with a waxlike waterproofing material that helps protect against desiccation.

CONSERVATION CONCERNS: None known.

DID YOU KNOW? *Xylocopa* eggs are probably the largest of all insect eggs. An egg can be more than half an inch (13 mm) long, which is more than half the length of the female's body. Each *Xylocopa* female lays only a small number of eggs in her lifetime (usually eight or fewer) and, compared to other bees, invests greater maternal care and more time in rearing them.

SUBFAMILY: Xylocopinae

Ceratina: Small Carpenter Bees

(ser-ah-TIE-nah)

0.1 TO 0.6 INCHES
(3–15 MM) LONG

FROM A DISTANCE, small carpenter bees look black, but up close they may be a handsome metallic green with bronzy overtones. The genus *Ceratina* is found on all continents except Antarctica with about 21 species in the United States and Canada. They are rare in desert habitats.

IDENTIFICATION: Small carpenter bees are sturdy, shiny, sparsely haired, and are black, blue, or green. Tiny to moderate, 0.1 to 0.6 inches (3–15 mm) long. The abdomen is a distinctive shape; it appears almost cylindrical with a rather blunt end. However, up close it's possible to see a tiny pointed tip. Most species have yellow or white markings on the face; males generally have larger pale markings (frequently in the shape of a broad inverted T). Females carry dry pollen on sparsely haired scopae on their hind legs.

SIMILAR INSECTS: May be confused with small, shiny halictid bees. A *Ceratina* can be distinguished from these by the lack of pollen hairs on its **femur** (the longer leg segment close to the bee's body) and its distinctive cylindrical abdomen.

FORAGING: Most species visit a wide variety of flowers for nectar and pollen.

NESTS: North American *Ceratina* are almost exclusively solitary. They excavate their nests but lack the jaw strength to chew solid wood, choosing instead the pithy centers of dead stems of sunflowers and shrubs such as elderberry, box elder, sumac, and blackberry. The stems must be broken to allow nesting *Ceratina* access to the pith.

The female makes a linear series of brood cells (using chewed pith and saliva to make dividing walls) and when finished, places herself as a guard at the entrance. She will die during the winter but remains in place to block the nest access. Her offspring complete development in the fall and remain in the nest as adults until spring.

CONSERVATION CONCERNS: None known.

DID YOU KNOW? This genus is unusual because it has several species that are *parthenogenetic:* females can produce female eggs without mating. One of the best known parthenogenetic species is *Ceratina dallatorreana*, a native of Europe that has been introduced to California.

SUBFAMILY: Nomadinae

Nomada: Cuckoo Bees

(no-MAH-duh)

0.1 TO 0.6 INCHES
(3–15 MM) LONG

AS WITH OTHER CUCKOO BEES (see *Sphecodes* on page 238, *Coelioxys* on page 245, and *Stelis* on page 244), *Nomada* females do not make their own nests but lay eggs in the nests of other species. The genus *Nomada* has approximately 290 species in the United States and Canada. They are common in temperate areas but rare in the tropics, and they can be found as far north as Alaska.

IDENTIFICATION: Slender, sparsely haired, wasplike, tiny to moderate-sized bees, 0.1 to 0.6 inches (3–15 mm) long. Most species are black or red, and the majority have yellow or white markings. The surface of the thorax is rough and covered with tiny pits. Their antennae appear thick in comparison to other bees. Because they do not collect pollen, *Nomada* lack pollen-carrying structures such as scopae or corbiculae.

SIMILAR INSECTS: Easily mistaken for small sphecoid wasps, *Nomada* can be distinguished by the hairs on their faces. The hairs on wasps are a single filament and thus reflect light, so those on the lower part of the face gleam silver or gold in light. *Nomada* in contrast, like all bees, have branched (feathery) hairs that do not reflect light, so their hairs appear dull.

FORAGING: *Nomada* females do not collect pollen; their larvae feed on the pollen of the host bees. Both males and females visit flowers for nectar.

NESTS: All species are cleptoparasites on other bees, primarily *Andrena* but also *Agapostemon*, *Halictus*, and *Lasioglossum*. In an *Andrena* nest, *Nomada* may lay two to four eggs alongside the one egg found in each host brood cell. In other host nests, *Nomada* lay the more typical single egg. Once emerged, the *Nomada* larva destroys the host egg (also its siblings' eggs in an *Andrena* nest). To accomplish this, first-stage *Nomada* larvae have large, sickle-shaped mouthparts, which are lost during the first molt. Subsequent larval instars possess normal mouth parts, which they use to consume the host cell's pollen provisions.

CONSERVATION CONCERNS: None known.

DID YOU KNOW? Cuckoo bees find host nests by smell. Male *Nomada* can mimic the odor of female *Andrena* and will patrol *Andrena* nest sites emitting the scent. This behavior may make it easier for *Nomada* females to find a nest site. The male also may transfer the scent to the female during mating, helping her mask her presence when entering the *Andrena* nest.

SUBFAMILY: Apinae

Eucera: Long-Horned Bees
(you-SIR-uh)

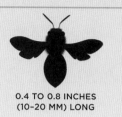

0.4 TO 0.8 INCHES
(10–20 MM) LONG

Eucerini, which includes *Melissodes, Svastra,* and the rare *Tetralonia.* They are all a similar shape, rather hairy, and often with pale hair bands. *Eucera* may also be confused with some *Diadasia.*

FORAGING: Little is known about the foraging behavior, but most species probably visit a wide variety of flowers, especially sunflowers.

NESTS: Most *Eucera* make solitary nests in the ground. Like many other ground-nesting bees, the female secretes a waterproof waxlike material to line the walls of her brood cells.

CONSERVATION CONCERNS: Two species are on the Xerces Society's Red List (see page 13). One, *Eucera frater lata,* is endemic to but widespread within the Columbia River basin. It prefers visiting flowers of milkvetches (*Astragalus*) and lives in a variety of habitats. Another species, *E. douglasiana,* is known from only a single specimen collected in Washington in 1902.

LIKE OTHER CLOSELY RELATED BEES such as *Melissodes,* males often have very long antennae, resulting in the common name long-horned bees. *Eucera* is a large genus with nearly 120 species in North America.

IDENTIFICATION: Small to moderate-sized bees, 0.4 to 0.8 inches (10–20 mm) long. The head and thorax of most *Eucera* are covered with dense, pale brown hair. Faces are often yellow, while abdomens are often dark colored and may have bands of pale hair. Females carry dry pollen in scopae, which cover the lower half of their hind legs.

SIMILAR INSECTS: *Eucera* is one of a group of closely related genera placed in the tribe

DID YOU KNOW? *Eucera* is one of a group of closely related genera placed in the tribe Eucerini, which includes *Melissodes, Svastra,* and *Tetralonia.* They are so much alike that some scientists have suggested combining them all into *Eucera* as a single genus. There is still debate over how these generally confusing bees should be classified.

SUBFAMILY: Apinae

Melissodes: Long-Horned Bees
(mel-ih-SO-deez)

**0.3 TO 0.7 INCHES
(7–18 MM) LONG**

LIKE OTHER CLOSE RELATIVES such as *Eucera*, these bees get their name from very long antennae found on males of most species. *Melissodes* is a large and common genus, with more than 120 species in North America.

IDENTIFICATION: Robust and hairy, small to moderate-sized bees, 0.3 to 0.7 inches (7–18 mm) long. They have conspicuously hairy rear legs, and usually have bands of pale hair on their abdomens. The female carries dry pollen in large scopae on the lower half of her hind legs. The lower front part of the males' faces are frequently yellow, and they often have very long antennae.

SIMILAR INSECTS: Very similar to *Svastra*, though those bees are usually larger, up to 0.8 inches (20 mm) long. Can also be confused with other bees in the tribe Eucerini, which includes *Eucera, Svastra,* and the rare *Tetralonia*. (See opposite page.)

FORAGING: *Melissodes* have a particular association with sunflowers, though many are aster or daisy specialists. They are important pollinators both of wild sunflowers and of hybrid sunflower crops. The strong association with sunflowers is reflected in the habits of males; they will patrol sunflowers for mates during the day and sleep in clusters on the flower heads. *Melissodes* fly from midsummer through fall, and they tend to be active during mornings and early afternoons.

NESTS: All *Melissodes* nest in the ground. Most are solitary, excavating nests in flat ground, but some dig horizontal burrows in exposed banks. A few species nest communally: several individuals share one nest entrance but provision their own brood cells. As with most other ground-nesting bees, females secrete a waxlike material to line their brood cells.

CONSERVATION CONCERNS: None known.

DID YOU KNOW? *Melissodes* are parasitized by cuckoo bees in the subgenus *Triepeolus*. The *Triepeolus* female enters the *Melissodes* nest and lays its own egg into the wall of a partially provisioned brood cell, where it remains hidden until the cell is completed. After hatching, the *Triepeolus* larva kills the host egg or larva and then eats the pollen provisions. The name Melissa comes from the Ancient Greek word for honey, though it has come to mean bee.

SUBFAMILY: Apinae

Anthophora: Digger Bees
(an-THOH-for-uh)

0.25 TO 1.0 INCH
(5–25 MM) LONG

THOUGH THEY ARE ALL ROBUST, hairy, and fast flying, the species of *Anthophora* vary greatly in size and colors. *Anthophora* is a large genus with more than 400 species found in temperate and tropical regions worldwide. There are about 70 species in the United States; they are most abundant and diverse in the western states.

IDENTIFICATION: Robust, hairy, small to large bees, 0.25 to 1.0 inch (5–25 mm) long. Many have bands of pale hair on the abdomen, and all females have obviously hairy hind legs thanks to their scopae. These fast-flying bees are often heard before they are seen.

SIMILAR INSECTS: *Centris* bees are very similar but have bulkier and denser scopae on their hind legs. (They are denser because they are for carrying oil, not pollen.) Some *Anthophora* (species in the subgenus *Melea*) apparently are mimics of bumble bees (genus *Bombus*). These can vary in color from one region to another.

FORAGING: *Anthophora* visit a wide variety of flowers for nectar and pollen. They have very long tongues that enable them to drink nectar in deep flowers. One species, *A. urbana*, has been shown to be an important pollinator of cherry tomato crops.

NESTS: Most dig nests into vertical banks or flat ground, preferring loam or sandy loam soils. They can even be found nesting in aggregations at the beach. A few species nest in hard clay (such as adobe), softening the soil with regurgitated nectar and water to ease digging. Nest entrances may or may not be marked by a mound of soil; species in the subgenus *Melea* nest in banks and construct a downward-curving tube as the entrance to each nest. The exception to ground nesting comes from species in the subgenus *Clisodon*, which dig nests in rotten wood or pithy stems.

Brood cells are usually vertical and lined with an oily substance that makes them almost completely waterproof. The food provisions are a wet mixture of pollen and nectar that fills the lower part of the cell. Most species overwinter in the brood cell, either in the last larval stage (prepupa) or a few as adults.

CONSERVATION CONCERNS: None known.

DID YOU KNOW? *Anthophora* males, like those of the closely related genera *Eucera* and *Melissodes*, often form sleeping aggregations of several individuals on a single plant. They fall asleep holding onto the plant with their jaws.

SUBFAMILY: Apinae

Centris: Mining Bees

(SEN-tris)

0.4 TO 1 INCH
(10-25 MM) LONG

WITH THE EXCEPTION of the largest species, *Centris* are adept fliers that can fly very fast and hover well, too. Many are unusual in that they collect plant oils instead of nectar. The oils are used to waterproof the brood cells; they probably use the oil in the place of nectar to provision the brood cells as well. The genus is restricted to the Americas. It has 85 species in North America and is most abundant and diverse in the tropics and deserts. It is rare elsewhere.

IDENTIFICATION: Compact, robust, and very hairy. Moderate to large bees, 0.4 to 1 inch (10–25 mm) long. The head and thorax of most species are covered with dense, pale hairs. The rest of the bee is black, although sometimes the abdomen is red or has a weak metallic green or blue sheen, and it often has yellow markings. A number of species have strikingly colored red or yellow eyes. To carry plant oils instead of pollen, *Centris* have very large bushy scopae on their hind legs. They often fly with a very loud hum.

SIMILAR INSECTS: *Anthophora* are similar, but the scopa on their rear legs appear smaller and are made of shorter and less dense hairs.

FORAGING: *Centris* have long tongues that enable them to extract provisions from deep tubular flowers; many species collect plant oils instead of nectar for the brood cell provision. They will collect pollen from a range of plants. In the desert Southwest, creosote bush (*Larrea*) and mesquite (*Prosopis*) are important sources.

NESTS: *Centris* are solitary nesting bees. The majority dig nests in loose, sandy soil, either in flat areas or in a bank. Nests in flat ground are usually fairly shallow, with vertical brood cells 3 to 10 inches (8–25 cm) deep. A few species make nests in preexisting holes in logs or branches to which they carry soil, so even here the nests are often made in or of soil.

CONSERVATION CONCERNS: *Centris errans* (Florida locustberry oil-collecting bee) is known from about 10 localities in the Florida Keys and extreme southern Florida Peninsula.

DID YOU KNOW? The males of most species establish and defend a mating territory. It is usually a few square yards in size, marked with scent from a gland near their mouths. This territory will be maintained for days at a time, the male chasing after almost anything that enters, even a tossed pebble. In a few species, males will dig into nests to mate with females before they emerge.

SUBFAMILY: Apinae

Habropoda: Digger Bees
(Hab-roh-PO-duh)

**0.4 TO 0.7 INCHES
(10–18 MM) LONG**

AROUND 15 SPECIES of *Habropoda* are found in the United States, with most occurring in the far west coast from British Columbia south to Baja California. Only one species is found east of Texas and the Great Plains, although it is a species of some economic importance.

IDENTIFICATION: *Habropoda* are moderate-sized bees, 0.4 to 0.7 inches (10–18 mm) in length. They are extremely round, robust, and hairy bees that are usually gray in color, occasionally with striped abdomens. A few species are other colors such as black or brown.

SIMILAR INSECTS: Many *Habropoda* resemble *Anthophora* in color, size, and body shape. One species, *H. laboriosa* has a black abdomen and tan thorax, giving it an appearance similar to that of bumble bees.

FORAGING: Little is known about the foraging preferences of most *Habropoda* species. The ability of some species to "buzz-pollinate" flowers by vibrating their flight muscles to shake pollen from the anthers has been documented.

NESTS: *Habropoda* are ground-nesting bees, with some species known to prefer sandy soils. One species, *H. miserabilis*, has been reported to construct only one cell at the end of each burrow.

CONSERVATION CONCERNS: None known.

DID YOU KNOW? The southeastern blueberry bee (*Habropoda laboriosa*) is a specialist forager of rabbiteye blueberry (*Vaccinium virgatum*) in the southeastern United States. Scientists have documented individual bees pollinating up to 6,000 berries during an average lifespan, a yield valued at more than $20.

SUBFAMILY: Apinae

Bombus: Bumble Bees
(BOM-bus)

0.4 TO 0.9 INCHES
(10-23 MM) LONG

BUMBLE BEES ARE among the most easily recognized and best-loved bees. There are nearly 50 species in North America, found primarily in temperate areas and ranging as far north as there is land.

IDENTIFICATION: Robust and very hairy, moderate to large bees, 0.4 to 0.9 inches (10–23 mm) long. They have yellow, black, white, brown, or orange bands by which different species can be distinguished, although there can be variation in color pattern even within a species. *Bombus* females carry pollen moistened with nectar in stiff hair baskets (corbiculae) on their hind legs.

SIMILAR INSECTS: Few bees are large enough to be confused with bumble bees. Large carpenter bees (*Xylocopa*) are generally not as hairy nor as brightly colored. *Centris* mining bees have distinctive, very large, hairy scopae on their rear legs. Digger bees (*Anthophora*) usually are not as brightly colored; a few species that mimic bumble bees may look like smaller bumble bee workers, although possessing scopae (an area of long, dense hairs on the hind leg) rather than the bumble bee's corbiculae (an open area surrounded by long, incurving hairs). In addition, several syrphid flies are excellent bumble bee mimics.

FORAGING: Bumble bees are among the first bees to emerge in the spring and the last to disappear in the fall. They visit a succession of flowers from a wide variety of plants throughout the foraging season, from early-flowering willows to late-summer blooms such as goldenrod. Many species have long tongues that enable them to access nectar from deep flowers such as larkspur

and penstemon. *Bombus* are important pollinators of crops as diverse as tomatoes, watermelons, and blueberries.

NESTS: *Bombus* nest socially in annual colonies. At the end of summer most bumble bees die, leaving only a few mated queens to hibernate. In spring, the queens emerge, each founding a new colony as a solitary bee. The queen rears her initial brood; once those workers are ready to take over foraging, she remains in the nest to lay eggs.

Bombus nests consist of an irregular cluster of ball-like, wax brood cells in a small cavity such as an abandoned rodent burrow or under a grass tussock. The cells are unique among all bees because they may contain multiple offspring and are enlarged as the larvae develop. Provisioning of the cells is progressive, meaning that additional provisions are added as required. Bumble bees store small quantities of nectar, enough to supply the colony for only a couple of days.

Bumble bees in the subgenus *Psithyrus* are social parasites of other bumble bees. A *Psithyrus*

(continued on next page)

Bombus: Bumble Bees
(continued)

female will enter an established nest and then usurp the host queen, either instantly by killing her or gradually by living alongside her. The *Psithyrus* queen is not always successful — she may be driven away or killed by the current occupants. Once established in the nest, she will lay her own eggs that the workers will tend. *Psithyrus* look similar to other bumble bees but lack pollen-gathering structures because they do not forage.

CONSERVATION CONCERNS: There has been an abrupt and steep decline in several bumble bee species, particularly *Bombus occidentalis*, *B. affinis*, *B. terricola*, and *B. franklini*. Sadly, *B. franklini* may be extinct. Habitat loss and pesticide use probably contributed to this decline, but the principal reason is likely disease. In the 1990s, American bumble bees were taken to Europe for breeding for the U.S. greenhouse tomato industry. They appear to have contracted diseases that they carried back when they were subsequently reimported; these spread to wild populations.

DID YOU KNOW? Bumble bees can buzz-pollinate. Some flowers such as tomatoes need to be vibrated to release the pollen (at roughly the frequency of a middle C musical note). Bumble bees do this by grabbing onto the flower and then vibrating their flight muscles without flapping their wings. There is an audible buzz as they do this, hence the name.

FAMILY: Apidae

SUBFAMILY: Apinae

Apis: Honey Bees
(A-pis)

THERE IS ONE SPECIES OF HONEY BEE in North America, the European honey bee (*Apis mellifera*). This bee is native to Europe, Africa, the Middle East, and Central Asia, but it has been introduced to all other parts of the world. It arrived in the Americas with European colonists in the early seventeenth century. Honey bee colonies are widely managed for honey production and crop pollination purposes.

IDENTIFICATION: Honey bees are moderately hairy, elongated bees with hairy eyes; moderate-sized, 0.4 to 0.6 inches (10–15 mm) long. They vary in color from black to amber brown with stripes on their abdomens. Females have corbiculae on their rear legs for carrying moistened pollen. There are striking morphological differences between female workers, the queen, and males. The queen is much larger than the others, appearing almost wasplike. Males have large eyes and congregate in areas where they fly together and mate with newly emerged queens.

The queens find these aggregations of males by following pheromone scent trails.

SIMILAR INSECTS: The drone fly (*Eristalis tenax*) is an effective mimic of honey bees, but it has only one pair of wings and short antennae. It, too, is a European species now established in North America.

FORAGING: These generalist foragers can, and do, visit a huge variety of flowers for nectar and pollen. However, since they did not evolve in North America, they are not specifically adapted to native New World plants. For example, they are inefficient pollinators of Nightshade-family flowers such as tomato that require vibrating to release pollen. They may rob the flowers of blueberry and alfalfa for nectar without coming into contact with the flower's reproductive parts.

NESTS: *Apis mellifera* nests socially in perennial colonies, hoarding honey to support the colony through periods when foraging is limited. Honey bees occupy relatively large cavities, such as hives or hollow trees, in which they make vertical combs of hexagonal cells using wax secreted from special glands. In addition to wax, hive spaces are filled with propolis, a mixture of plant resins.

A. mellifera has morphologically distinct castes including the queen, male drones, and female workers. The queen's primary role is egg laying. Male drones exist only to fertilize the queen. Worker bees perform different tasks according to age. Young workers attend to the larvae, clean the nest, and care for the queen. Slightly older bees protect the nest entrance. The oldest workers work outside the hive foraging for food. New colonies form by division; a new queen is raised and the old one leaves the hive with a swarm of workers to establish a new nest.

CONSERVATION CONCERNS: Since the middle of the twentieth century, the number of managed honey bee colonies in the United States has halved, primarily due to the introduction of exotic parasites and diseases. In 2006, a new problem, Colony Collapse Disorder, emerged. The CCD name refers to its ability to wipe out entire bee colonies in a very short period of time. CCD is characterized by once busy hives found empty of bees, yet containing honey stores.

In the past few decades, feral honey bees descended from an African subspecies of the European honey bee (*Apis mellifera scutellata*) have spread into southern states from Mexico. These "Africanized" bees are the result of hybridization between various European honey bee races and an African race, which escaped from a bee breeding research center in Brazil. Africanized bees have gained a reputation for aggressive behavior and are colloquially known as "killer bees." As hybridization of these feral bees continues, however, the actual level of aggressiveness varies greatly between colonies.

DID YOU KNOW? *Apis mellifera* workers actively control the temperature of their hives, maintaining brood cells between 94.1 and 95.9°F for optimal development. Worker bees generate heat by shivering their flight muscles; they cool the hive during hot summer days by collecting water and fanning air into the nest.

PART 4

Creating a Pollinator-Friendly Landscape

Pollinator-watching, like bird-watching, provides a unique opportunity to connect with nature in virtually any environment — from expansive western prairies to inner-city rooftops. Indeed, many of the world's most notable invertebrate ecologists, such as E. O. Wilson, Bert Hölldobler, Tom Eisner, and Robert Michael Pyle, recount transformative childhood moments observing insects in vacant lots and on the edges of city sidewalks.

With fairly minimal efforts, landscapes from farms to urban spaces, school yards to golf courses can be enhanced to support diverse insect communities, bringing the enjoyment of pollinator watching to young and old alike. Simple decisions about selecting plants, providing nest sites, minimizing disturbance, and reducing pesticides can make a dramatic difference between a green, manicured, but lifeless landscape, and one that teems with the color, energy, and life of buzz-pollinating bumble bees, rapidly dashing hummingbird moths, and busy nest-building leafcutter bees.

SAMPLE GARDENS

The following plans are offered to help you envision a transformation of your landscape into one that welcomes pollinators, whether it be a suburban flower bed, a working farm, an inner-city community vegetable garden, a roadside planting, or a public space.

Using These Plans

The landscape plans in this section should be approached for inspiration rather than exact replication. Indeed, specific local conditions, such as soil moisture, sun exposure, and human utility, will dictate the placement and selection of wildflowers and woody plants.

There are, however, a few basic principles that these landscape designs share. First, they all attempt to maintain open, unshaded conditions. While some trees are excellent pollen and nectar sources, most bees and butterflies prefer sunny landscapes and meadowlike conditions. Trees are often best along property edges or the north sides of buildings where wildflowers are difficult to grow.

The second feature all of these designs share is a diversity of flowers that bloom throughout the growing season. Each of these designs shows at least three different flower species in bloom. This minimum standard should be maintained for each of the seasons — spring, summer, and fall — as well as winter if you live in a warm climate. This three-species-per-season benchmark should be considered a minimum rather than an end goal.

Finally, each of these diagrams demonstrates a landscape rich in color and reduced of grasses. Pollinators are most attracted to large blocks of color and forage most efficiently where the same flower species are clustered together. With that in mind, consider grouping single species together for maximum effect.

Compare these illustrations with your own landscape. What features do they share? Where can you place taller plants so that they will not shade out smaller ones? Where does water drain, and how might that affect the placement of moisture-loving plants? How can you use plants to accentuate vistas or hide eyesores? And where might you place plants to maintain human safety and utility?

> **PLANT RECOMMENDATIONS**
>
> For regional plant lists, including trees, shrubs, and tall and short wildflowers, see pages 272–279. For a fuller plant-by-plant listing, see pages 284–323.

KEY

A TREES
B SHRUBS
C TALL WILDFLOWERS (>4')
D VEGETABLES
E NEST BLOCK

S etting aside part of a community garden specifically for pollinators provides direct benefits in the form of larger, more abundant, and better formed fruits and vegetables. Because native pollinators are generally docile and rarely sting, they are also a perfect way to connect urban dwellers to the natural world.

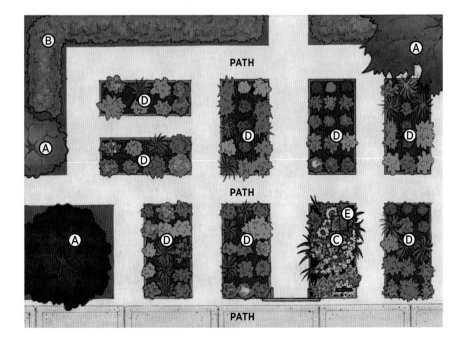

armers who want to take a more active role in increasing populations of resident native bees can increase the available foraging habitat to include a range of plants that bloom and provide abundant pollen and nectar from spring to fall.

KEY

A TREES
B STONE WALL
C MIXED WILDFLOWERS

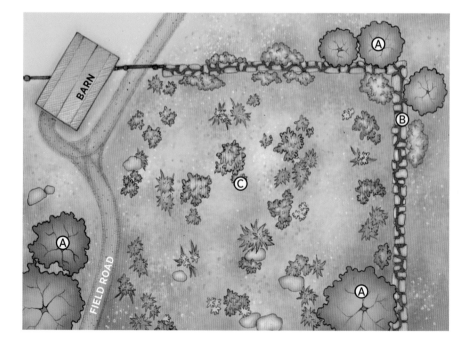

KEY

A TREES
B SHRUBS
C TALL WILDFLOWERS (>4')
D SHORT WILDFLOWERS (<4')
E NEST BLOCK

Flower beds can be designed and planted with a range of native flowers and heirloom varieties that provide nectar and pollen for bees, or are the preferred caterpillar host plants for butterflies. In fact, designing gardens so there's something blooming at all times is a classic aesthetic goal.

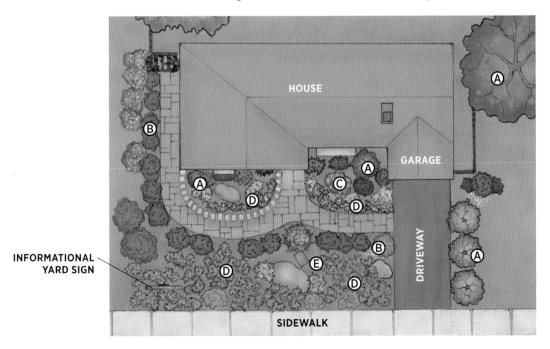

Long, narrow landscape features such as drainage ditches, fencerows, hedgerows, windbreaks, stream corridors, and roadsides can provide a corridor of habitat that links areas of a farm.

KEY

A TREES
B SHRUBS
C TALL WILDFLOWERS (>4')
D SHORT WILDFLOWERS (<4')
E NEST BLOCK
F CROPS

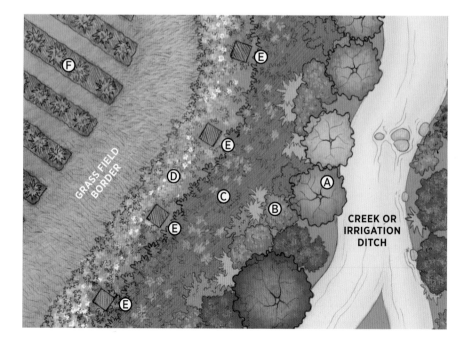

KEY

A TREES
B SHRUBS
C TALL WILDFLOWERS (>4')
D SHORT WILDFLOWERS (<4')

Roadside ditches are often the only semiwild habitat in intensively farmed or developed areas. Long-linear pathways can connect fragmented habitat, and once established, native wildflowers can crowd out tree seedlings and reduce mowing costs. Contrary to popular perception, roadside flowers don't usually increase traffic mortality of pollinators.

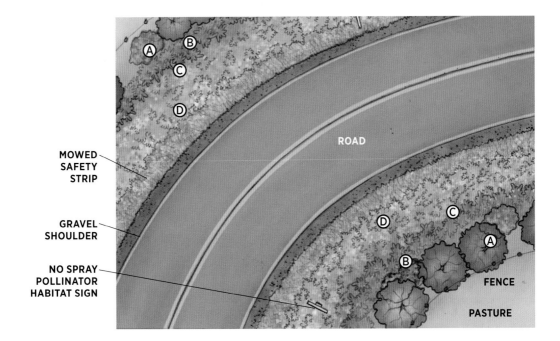

Pollinator gardens can highlight a school or business's environmental stewardship and provide a beautiful, relaxing, and educational setting for breaks, outdoor meetings, and classes. Native landscaping can also help meet various green-building certification standards.

KEY

A TREES
B SHRUBS
C TALL WILDFLOWERS (>4')
D SHORT WILDFLOWERS (<4')

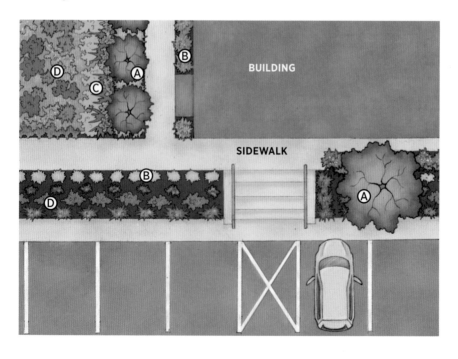

KEY

A TREES
B SHRUBS
C TALL WILDFLOWERS (>4')
D SHORT WILDFLOWERS (<4')
E NEST BOX
F NATIVE BEES NEST WALL

Public gardens can easily incorporate pollinators into their educational mission by featuring native bee nests, erecting interpretive signage, and highlighting excellent local pollinator plants. Tours and classes can point out the features of these demonstration gardens and inspire visitors to enhance their own home landscapes.

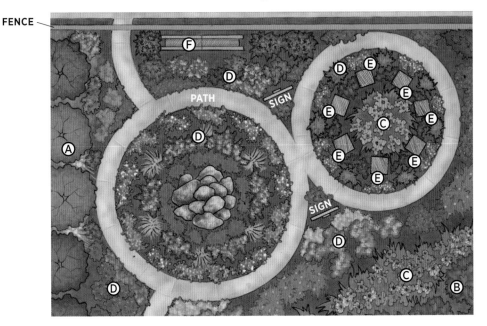

Midwestern United States

COMMON NAME	SCIENTIFIC NAME	BLOOM TIME	FLOWER COLOR	CULTURAL REQUIRE-MENTS
TREES				
American basswood	*Tilia americana*	Early summer	White	
SHRUBS				
Pussy willow	*Salix discolor*	Spring	Yellow	
Prairie rose	*Rosa arkansana*	Early summer	Pink	
Leadplant	*Amorpha canescens*	Summer	Purple	
TALLER WILDFLOWERS (OVER 4′)				
Wild bergamot	*Monarda fistulosa*	Summer	Light blue	
Purple giant hyssop	*Agastache scrophulariifolia*	Summer	White	
Prairie blazing star	*Liatris pycnostachya*	Summer	Lavender	
New England aster	*Symphyotrichum novae-angliae*	Fall	Purple	
Showy goldenrod	*Solidago speciosa*	Fall	Yellow	
SHORTER WILDFLOWERS (UNDER 4′)				
Wild lupine	*Lupinus perennis*	Spring	Blue	
Eastern waterleaf	*Hydrophyllum virginianum*	Spring	Pink	Shade
Spotted geranium	*Geranium maculatum*	Spring	Pink	Part shade
Smooth penstemon	*Penstemon digitalis*	Early summer	White	
Butterfly milkweed	*Asclepias tuberosa*	Summer	Orange	
Purple prairie clover	*Dalea purpurea*	Summer	Purple	
Eastern purple coneflower	*Echinacea purpurea*	Summer	Purple	
Riddell's goldenrod	*Solidago riddellii*	Fall	Yellow	

Great Plains and Prairie Provinces

COMMON NAME	SCIENTIFIC NAME	BLOOM TIME	FLOWER COLOR	CULTURAL REQUIRE-MENTS
TREES				
Chokecherry	*Prunus virginiana*	Spring	White	
SHRUBS				
Saskatoon serviceberry	*Amelanchier alnifolia*	Spring	White	
Pussy willow	*Salix discolor*	Spring	Yellow	
Prairie rose	*Rosa arkansana*	Early summer	Pink	
Leadplant	*Amorpha canescens*	Summer	Purple	
TALLER WILDFLOWERS (OVER 4')				
White wild indigo	*Baptisia alba*	Spring	White	
Wild bergamot	*Monarda fistulosa*	Summer	Light blue	
Showy milkweed	*Asclepias speciosa*	Summer	Pink	
Compassplant	*Silphium laciniatum*	Summer	Yellow	
Rough blazing star	*Liatris aspera*	Summer	Lavender	
Smooth blue aster	*Symphyotrichum laeve*	Fall	Blue	
Showy goldenrod	*Solidago speciosa*	Fall	Yellow	
SHORTER WILDFLOWERS (UNDER 4')				
Prairie spiderwort	*Tradescantia occidentalis*	Late spring	Blue	
Largeflowered beardtongue	*Penstemon grandiflorus*	Early summer	Pink	
Purple prairie clover	*Dalea purpurea*	Summer	Purple	
Narrow leaved coneflower	*Echinacea angustifolia*	Summer	Purple	
Zigzag goldenrod	*Solidago flexicaulis*	Fall	Yellow	
White heath aster	*Symphyotrichum ericoides*	Fall	White	

Pacific Northwest and British Columbia

COMMON NAME	SCIENTIFIC NAME	BLOOM TIME	FLOWER COLOR	CULTURAL REQUIRE- MENTS
TREES				
Manzanita	*Arctostaphylos* spp.	Spring	White/pink	
Pacific madrone	*Arbutus menziesii*	Spring	White	
SHRUBS				
Oceanspray	*Holodiscus discolor*	Summer	White	
Oregon grape	*Mahonia aquifolium*	Spring	Yellow	
TALLER WILDFLOWERS (OVER 4′)				
Showy milkweed	*Asclepias speciosa*	Summer	Pink	
Maximilian sunflower	*Helianthus maximiliani*	Summer	Yellow	
Wild buckwheat	*Eriogonum* spp.	Late summer	White/ pink/ yellow	
Canada goldenrod	*Solidago canadensis*	Fall	Yellow	
SHORTER WILDFLOWERS (UNDER 4′)				
California poppy	*Eschscholzia* spp.	Spring	Yellow	
Ithuriel's spear	*Triteleia laxa*	Spring	Blue/ purple	
Lupine	*Lupinus* spp.	Spring	Blue	
Perennial blanketflower	*Gaillardia aristata*	Summer	Orange	
Venus penstemon	*Penstemon venustus*	Summer	Pink	
Yellow rabbitbrush	*Chrysothamnus viscidiflorus*	Fall	Yellow	
Western mountain aster	*Symphyotrichum spathulatum*	Fall	Pink	

Lower-Cost Ornamentals for Many Regions

COMMON NAME	SCIENTIFIC NAME	BLOOM TIME	FLOWER COLOR	CULTURAL REQUIRE- MENTS
TREES				
Crab apple	*Malus* spp.	Spring	Pink	Avoid sterile varieties.
SHRUBS				
Shrub rose	*Rosa rugosa*	Early summer	Pink	
Lavender	*Lavandula* spp.	Summer	Blue	
TALLER FLOWERS (OVER 4')				
Catmint/catnip	*Nepeta* spp.	Summer	White/blue	
Russian sage	*Perovskia atriplicifolia*	Summer	Blue	
Cosmos	*Cosmos bipinnatus*	Summer	Pink	
Common sunflower	*Helianthus annuus*	Summer	Yellow	Avoid "pollenless" varieties.
SHORTER FLOWERS (UNDER 4')				
Borage	*Borago officinalis*	Summer	Blue	
Oregano	*Origanum* spp.	Summer	Pink	
Showy stonecrop	*Hylotelephium spectabile*	Fall	Yellow	

POLLINATOR PLANTS

It is worth noting that there are many different pollinator plant lists available from numerous conservation organizations and agencies. Some of these are becoming dated as new studies deliver improved knowledge about plant preferences of pollinators. The plants included in this list have been documented, to the greatest extent possible, to be highly attractive to pollinators, especially native bees. These preferences have been documented by bee researchers like Julianna Tuell and Rufus Isaacs at Michigan State University, and in field surveys conducted by Xerces Society scientists.

This information has been backed up by data on honey bee foraging preferences. Because honey bees have been studied for more than a century, there is a wealth of information, including whole books on their preferred pollen and nectar sources. And because honey bees often visit the same types of flowers as native bees, that research helped guide the development of this list.

▽ *Time spent selecting the best plants for a site will reap great rewards in both aesthetic value and pollinator benefit.*

Selection Criteria

Various other selection criteria also shaped the development of this list, such as plant distribution, vigor, and growth habit. The list avoids rare or endangered plants, species that do not grow vigorously, or species that have specialized cultural requirements. In addition to more difficult establishment, rare species may outcross with locally at-risk populations.

The list prioritizes species according to flowering abundance. For example, shrubs that propagate vegetatively by shoots or suckers often produce fewer flowers and are less beneficial to pollinators than others.

Plants with undesirable cultural characteristics have been excluded. This typically includes weedy species, and species that act as alternate pest and disease hosts for commercial crops. There are exceptions to this and some species that are perfectly acceptable in one landscape may become a nuisance in another. For example, black locust (*Robinia pseudoacacia*) is a tough native tree that can be a great pollinator plant in harsh planting situations such as polluted industrial sites or areas prone to spray from de-icing salts. Its ability to spread rapidly may be less desirable in other landscapes, however, such as golf courses and suburban woodlots. It is important to consider your own landscape priorities and limitations when reviewing this list for ideas.

Finally, one of the underlying principals is that the plants included on this list should be commercially available either through local nurseries or as seed from various mail-order native plant seed producers. There are a few instances in which the authors have included plants that are of particularly high-value to pollinators and yet are not widely available. It is our hope that mention of them here will help build a market for those species and make them more accessible.

California poppy
(*Eschscholzia* spp.)

EXPOSURE: Sun to part shade

SOIL MOISTURE: Dry

BLOOM TIME: Late spring

FLOWER COLOR: Yellow, orange

MAXIMUM HEIGHT: 1 foot (30 cm)

REGIONS: California, Southwest, Pacific
Northwest; introduced elsewhere

Annual and short-lived perennials; all attract a diversity of bee species. The petals on *Eschscholzia californica* close at night, and then open again the following morning. The petals also may close during inclement weather. California poppy is generally drought tolerant and easily self-sows. It grows well in disturbed areas, following fires, and along roadsides and railroad rights-of-way.

Common deerweed
(*Lotus scoparius*)

EXPOSURE: Sun

SOIL MOISTURE: Dry

BLOOM TIME: Spring to summer

FLOWER COLOR: Yellow

MAXIMUM HEIGHT: 4 feet (1.2 m)

REGIONS: California and Arizona

BUTTERFLY HOST PLANT: Various blues

Thrives in dry scrub and chaparral areas, especially areas newly cleared by fire. Often found growing with California sagebrush and toyon, another excellent bee plant.

Culver's root
(*Veronicastrum virginicum*)

EXPOSURE: Sun

SOIL MOISTURE: Average to wet

BLOOM TIME: Summer

FLOWER COLOR: White

MAXIMUM HEIGHT: 5 feet (1.5 m)

REGIONS: U.S. and Canada from Manitoba
south to Texas and east to New England
and Florida

Flowers are very attractive to many bees and some butterflies, but blooms are short-lived. Culver's root does not thrive in dry sites or shady locations.

Eryngo (*Eryngium* spp.)

EXPOSURE: Sun to part shade

SOIL MOISTURE: Wet to dry

BLOOM TIME: Summer

FLOWER COLOR: White, blue

MAXIMUM HEIGHT: 6 feet (1.8 m)

REGIONS: Nationwide U.S. and southern Canada

MOTH HOST PLANT: Rattlesnake master borer moth (*Papaipema eryngii*)

A group of plants with white or pale blue globelike blossoms that attract an abundance of small bees and syrphid flies. Rattlesnake master, *Eryngium yuccifolium*, is the common commercially available species. The thick hollow stems of rattlesnake master are slow to break down and provide nest sites for tunnel-nesting bees.

Figwort (*Scrophularia* spp.)

EXPOSURE: Part shade

SOIL MOISTURE: Average

BLOOM TIME: Late spring to summer

FLOWER COLOR: Green, red

MAXIMUM HEIGHT: 6 feet (1.8 m)

REGIONS: Throughout the U.S., north to central Canada

Both carpenter's square (*Scrophularia marilandica*), limited to the eastern United States and Canada, and the more widely distributed lanceleaf figwort (*S. lanceolata*) are prolific nectar producers. They often leave clothes and skin wet with sticky nectar when one walks through a solid stand. Not surprisingly, these plants draw huge numbers of bees, wasps, flies, and hummingbirds, especially when planted in large clusters. Because of their nondescript flowers, figworts are difficult to find through commercial vendors.

Fireweed
(*Chamerion angustifolium*)

EXPOSURE: Sun to part shade

SOIL MOISTURE: Wet

BLOOM TIME: Summer

FLOWER COLOR: Pink

MAXIMUM HEIGHT: 5 feet (1.5 m)

REGIONS: Throughout the U.S. and Canada, except the South and southeastern U.S.

Restricted to cool climates and high altitudes, fireweed thrives in moist areas where woody plants have been removed by fire. Seed can remain viable in the soil for extended periods, then rapidly germinate when forest clearings occur. Visited primarily by bees.

Giant hyssop
(*Agastache* spp.)

EXPOSURE: Sun to part shade

SOIL MOISTURE: Dry to average

BLOOM TIME: Summer

FLOWER COLOR: White, purple, yellow, blue

MAXIMUM HEIGHT: 6 feet (1.8 m)

REGIONS: Nationwide U.S. and Canada except for the Southeast from Louisiana to Florida

Recommended species include anise hyssop (*Agastache foeniculum*) and purple giant hyssop (*A. scrophulariifolia*). All species are visited by a diverse variety of bees, butterflies, and occasionally hummingbirds. A small black sweat bee, *Dufourea monardae,* is a specialist of giant hyssop and beebalm in the Midwest.

Goldenrod (*Solidago* spp.)

EXPOSURE: Sun to part shade

SOIL MOISTURE: Wet to dry

BLOOM TIME: Autumn

FLOWER COLOR: Yellow, white

MAXIMUM HEIGHT: 4 feet (1.2 m)

REGIONS: Nationwide U.S. and Canada

Some recommended species include showy goldenrod (*Solidago speciosa*), Riddell's goldenrod (*S. riddellii*), and stiff goldenrod (*S. rigida*). Goldenrods, where they exist, are among the most important late-season pollinator plants. Honey bees frequently use goldenrod to collect large amounts of nectar prior to winter; other bees use the pollen to provision late-season nests. Many beneficial solitary wasps are also frequent goldenrod visitors, as are pollen-eating beetles such as the soldier beetle (*Chauliognathus pennsylvanicus*) and the black blister beetle (*Epicauta pennsylvanica*). Goldenrods, along with asters, are also believed to be preferred floral sources for *Andrena hirticincta*, *A. nubecula*, *A. placata*, *A. simplex*, *A. solidaginis*, *Colletes simulans armatus*, and *Melissodes druriella*, all oligolectic bees of these Aster-family flowers.

Hayfield tarweed (*Hemizonia congesta*)

EXPOSURE: Sun

SOIL MOISTURE: Average to dry

BLOOM TIME: Autumn

FLOWER COLOR: White, yellow

MAXIMUM HEIGHT: 3 feet (0.9 m)

REGIONS: California and Oregon

An annual species with spindly, thin stems, and a subtle tarlike odor to the foliage. The plants produce daisy-like flower heads. Common flower visitors include small bees and various fly species.

Ironweed (*Vernonia* spp.)

EXPOSURE: Sun

SOIL MOISTURE: Average to wet

BLOOM TIME: Summer

FLOWER COLOR: Purple

MAXIMUM HEIGHT: 6 feet (1.8 m)

REGIONS: Western, midwestern, and southern U.S. north to Manitoba and Ontario

Ironweed blooms are short-lived but large and attract many butterflies and bees, including the specialist bees *Melissodes denticulata* and *M. vernoniae*. Recommended species include prairie ironweed (*Vernonia fasciculata*), giant ironweed (*V. gigantea*), and Missouri ironweed (*V. missurica*).

Ithuriel's spear
(*Triteleia laxa*)

EXPOSURE: Sun

SOIL MOISTURE: Dry

BLOOM TIME: Early summer

FLOWER COLOR: Blue, purple

MAXIMUM HEIGHT: 3 feet (0.9 m)

REGIONS: California and Oregon

Ithuriel's spear is a native plant that is increasingly available as a horticultural garden variety. It grows from a corm, which can be ordered from many bulb suppliers. Avoid planting in wet areas to avoid root rot.

Joe-Pye weed & Boneset
(*Eupatorium* spp.)

EXPOSURE: Sun to part shade

SOIL MOISTURE: Average to wet

BLOOM TIME: Summer

FLOWER COLOR: Pink or purple

MAXIMUM HEIGHT: 6 feet (1.8 m)

REGIONS: Central Canada south to Texas and east to the Atlantic coast from Florida north to Quebec

Joe-Pye weed is primarily known as a butterfly plant, but it also supports many solitary bees, bumble bees, and other insects. Recommended species include spotted Joe-Pye weed (*Eupatorium maculatum*) and sweetscented Joe-Pye weed (*E. purpureum*).

Lobelia (*Lobelia* spp.)

EXPOSURE: Sun to part shade
SOIL MOISTURE: Average to wet
BLOOM TIME: Summer
FLOWER COLOR: Red, blue, white
MAXIMUM HEIGHT: 6 feet (1.8 m)
REGIONS: Nationwide U.S. and Canada

Recommended species include cardinalflower (*Lobelia cardinalis*), a red-flowered plant pollinated by hummingbirds, and great blue lobelia (*L. siphilitica*), an exceptional bumble bee plant.

Lupine (*Lupinus* spp.)

EXPOSURE: Sun to part shade
SOIL MOISTURE: Average
BLOOM TIME: Spring
FLOWER COLOR: Blue, purple, pink, white
MAXIMUM HEIGHT: 6 feet (1.8 m)
REGIONS: Nationwide U.S. and Canada
BUTTERFLY HOST PLANT: Various blues, including the Karner blue; painted lady

Both perennial and annual species are available. Long-tongued bumble bees are the most common flower visitors, along with some mason bees and occasionally honey bees.

Milkweed (*Asclepias* spp.)

EXPOSURE: Sun
SOIL MOISTURE: Wet to dry
BLOOM TIME: Summer
FLOWER COLOR: White, purple, pink, orange
MAXIMUM HEIGHT: 5 feet (1.5 m)
REGIONS: Nationwide U.S. and Canada
BUTTERFLY AND MOTH HOST PLANT: Monarch, milkweed tussock moth

Milkweed flowers are visited by many bees, wasps, flies, butterflies such as swallowtails and fritillaries, and even hummingbirds. Many locally adapted species exist throughout North America, including the showy butterfly milkweed (*Asclepias tuberosa*) and swamp milkweed (*A. incarnata*).

Mountainmint
(*Pycnanthemum* spp.)

EXPOSURE: Sun

SOIL MOISTURE: Average to dry

BLOOM TIME: Late summer

FLOWER COLOR: White

MAXIMUM HEIGHT: 4 feet (1.2 m)

REGIONS: Central and southern U.S. north to eastern Canada

Virginia mountain mint (*Pycnanthemum virginianum*) is the most widely available species, but roughly a dozen other locally adapted species can be found in eastern North America. Most have a mintlike odor when the foliage is crushed.

Prairie clover (*Dalea* spp.)

EXPOSURE: Sun

SOIL MOISTURE: Average to dry

BLOOM TIME: Summer

FLOWER COLOR: White, purple

MAXIMUM HEIGHT: 3 feet (0.9 m)

REGIONS: Nationwide U.S. except New England; southern Canada

BUTTERFLY HOST PLANT: Various sulfurs, Isola blue (*Hemiagus isola*)

Purple prairie clover (*Dalea purpurea*) is a widespread perennial across much of North America and is one of the most attractive summer-blooming bee plants, with prolific purple flowers. It should be considered a "must-have" in pollinator conservation efforts where it is native and appropriate to the site. Four reported specialist pollinators of these plants include the polyester bees *Colletes albescens, C. susannae, C. wilmattae,* and *C. robertsonii*. Prairie clovers are also voraciously visited by honey bees and bumble bees.

Purple coneflower
(*Echinacea* spp.)

EXPOSURE: Sun

SOIL MOISTURE: Average

BLOOM TIME: Summer

FLOWER COLOR: Purple

MAXIMUM HEIGHT: 2 feet (0.6 m)

REGIONS: Midwestern and eastern U.S.

BUTTERFLY HOST PLANT: Various checker-spots

Pale purple coneflower (*Echinacea pallida*) and narrow-leaved coneflower (*E. angustifolia*) attract many pollinators, but common purple coneflower (*E. purpurea*) is more adaptable and commercially available. All purple coneflowers are excellent bee and butterfly plants. Common visitors include bumble bees; sweat bees; bees in the genera *Diadasia*, *Melissodes*, and *Svastra*; the sunflower leafcutter bee (*Megachile pugnata*); and many butterflies such as monarchs, swallowtails, sulfurs, and various others. *Andrena helianthiformis* is a specialist bee of purple coneflower.

Rod wirelettuce
(*Stephanomeria virgata*)

EXPOSURE: Sun

SOIL MOISTURE: Dry

BLOOM TIME: Autumn

FLOWER COLOR: White, purple

MAXIMUM HEIGHT: 6 feet (1.8 m)

REGIONS: California, Oregon, Nevada

An annual adapted to dry sites, with multiple flowers arising from a central stalk. Rod wirelettuce is common on open, dry slopes ranging from coastal sage scrublands inland to yellow pine forests. While not considered to be the most attractive native plant, rod wirelettuce is appreciated for its ability to grow in dry, exposed locations where few other pollinator plants thrive.

Rosinweed (*Silphium* spp.)

EXPOSURE: Sun

SOIL MOISTURE: Average to dry

BLOOM TIME: Summer

FLOWER COLOR: Yellow

MAXIMUM HEIGHT: 8 feet (2.4 m)

REGIONS: Central U.S., east to Florida and
north to Quebec

MOTH HOST PLANT: Silphium moth
(*Tabenna silphiella*)

These large, sunflower-like perennials attract primarily bees and soldier beetles. Cup plant (*Silphium perfoliatum*) has larger, more attractive flowers than other species, but rosinweed (*S. integrifolium*) and compassplant (*S. laciniatum*) are also excellent choices. All are tough, long-lived, and have extremely deep root systems. Cup plant has thick hollow stems that decompose slowly over several seasons. It is not uncommon for broken stems to serve as nest sites for *Ceratina* carpenter bees and various leafcutter bees.

Scorpionweed (*Phacelia* spp.)

EXPOSURE: Sun to part shade

SOIL MOISTURE: Average to dry

BLOOM TIME: Spring

FLOWER COLOR: White, purple, yellow, pink

MAXIMUM HEIGHT: 3 feet (0.9 m)

REGIONS: Nationwide U.S. and Canada
except Florida, northern New England,
and eastern Canada

Many annual and perennial Phacelia species are distributed across North America; the best are native to the western United States, although they have been introduced widely. Recommended plants include lacy phacelia (*Phacelia tanacetifolia*), California phacelia (*P. californica*), imbricate phacelia (*P. imbricata*), and largeflower phacelia (*P. grandiflora*). Some specialist bees of *Phacelia* include *Andrena lamelliterga* and *A. phaceliae*.

Sneezeweed (*Helenium* spp.)

EXPOSURE: Sun
SOIL MOISTURE: Average to wet
BLOOM TIME: Summer to autumn
FLOWER COLOR: Yellow
MAXIMUM HEIGHT: 3 feet (0.9 m)
REGIONS: Nationwide U.S. and Canada

Common sneezeweed (*Helenium autumnale*) is a widely distributed and commonly available plant that prefers wet sites. It attracts many bee species including honey bees, bumble bees, and leafcutter bees.

Spiderflower (*Cleome* spp.)

EXPOSURE: Sun
SOIL MOISTURE: Average to dry
BLOOM TIME: Summer
FLOWER COLOR: Yellow, white, red, pink
MAXIMUM HEIGHT: 3 feet (0.9 m)
REGIONS: Nationwide U.S. and Canada

Most spiderflowers are annuals, including yellow spiderflower (*Cleome lutea*) and Rocky Mountain beeplant (*C. serrulata*), which may have red, white, or pink flowers. The nonnative ornamental, *C. hassleriana*, seems to attract the fewest pollinators.

Spiderwort (*Tradescantia* spp.)

EXPOSURE: Sun to part shade
SOIL MOISTURE: Average
BLOOM TIME: Late spring to early summer
FLOWER COLOR: Blue, purple
MAXIMUM HEIGHT: 2 feet (0.6 m)
REGIONS: Southern Canada and nationwide U.S. except the Pacific Northwest

Commonly available and widely distributed species include Ohio spiderwort (*Tradescantia ohiensis*), prairie spiderwort (*T. occidentalis*), and Virginia spiderwort (*T. virginiana*). Flowers on most species open in the early morning, then wilt by midday. Bumble bees and honey bees are common visitors.

Sunflower (*Helianthus* spp.)

EXPOSURE: Sun to shade

SOIL MOISTURE: Average to dry

BLOOM TIME: Late summer to autumn

FLOWER COLOR: Yellow, orange

MAXIMUM HEIGHT: 8 feet (2.4 m)

REGIONS: Nationwide U.S. and Canada

BUTTERFLY HOST PLANT: Various checker-spots

Many species and horticultural varieties are commonly available, including woodland sunflower (*Helianthus divaricatus*), suited to shady locations in the eastern United States; Maximilian sunflower (*H. maximiliani*), a tall species suited to dry sunny sites; and prairie sunflower (*H. petiolaris*). All species attract a tremendous diversity of insects including bees, wasps, flies, butterflies, and pollen-eating beetles. When planting the common annual sunflower (*H. annuus*), avoid "pollenless" or double-petaled ornamental varieties. Bees in the genera *Diadasia*, *Melissodes*, *Eucera*, and *Svastra* are all common sunflower specialists. Other specialist bees of sunflower include *Andrena accepta*, *A. aliciae*, *A. helianthi*, *Dufourea marginatus*, *Melissodes agilis*, *Pseudopanurgus rugosus*, *Dieunomia heteropoda* (the largest member of the Halictidae bee family in the eastern United States), and the sunflower leafcutter bee (*Megachile pugnata*), which has been studied as a managed crop pollinator for sunflowers.

Vinegarweed
(*Trichostema lanceolatum*)

EXPOSURE: Sun

SOIL MOISTURE: Average to dry

BLOOM TIME: Autumn

FLOWER COLOR: Blue

MAXIMUM HEIGHT: 3 feet (0.9 m)

REGIONS: California

Vinegarweed is an annual mintlike herb, well adapted to dry summers and nutrient-poor soils. The common name is derived from a vinegar-like odor produced by crushing the foliage. Ornamental varieties of the closely related wooly bluecurls (*Trichostema lanatum*) are also available and are also excellent pollinator plants.

Waterleaf
(*Hydrophyllum* spp.)

EXPOSURE: Shade

SOIL MOISTURE: Average

BLOOM TIME: Spring

FLOWER COLOR: Purple, pink, white

MAXIMUM HEIGHT: 1 foot (30 cm)

REGIONS: Nationwide U.S. except the Gulf Coast, north to southern Canada

In most species, flower clusters rise from long stalks; triangular leaves are dappled by lighter-colored marks as though the leaves are stained by water. Bumble bee queens are common spring visitors. The blue orchard bee (*Osmia lignaria*) has been described as a common visitor, as has *Andrena geranii* and the waterleaf cuckoo bee (*Nomada hydrophylli*), both specialists of eastern waterleaf (*Hydrophyllum virginianum*).

Western rosinweed
(*Calycadenia* spp.)

EXPOSURE: Sun

SOIL MOISTURE: Dry

BLOOM TIME: Summer to autumn

FLOWER COLOR: White, yellow

MAXIMUM HEIGHT: 3 feet (0.9 m)

REGIONS: California, Nevada

These closely related annual plants include Fremont's western rosinweed (*Calycadenia fremontii*), Oregon western rosinweed (*C. truncata*), and several others. Western rosinweeds are members of the Aster family and generally have open, dish-shaped flowers that are perfect for small bees and various fly species.

Wild buckwheat
(*Eriogonum* spp.)

EXPOSURE: Sun

SOIL MOISTURE: Dry

BLOOM TIME: Summer to autumn

FLOWER COLOR: White, yellow, pink

MAXIMUM HEIGHT: 6 feet (1.8 m)

REGIONS: Nationwide U.S. and Canada except the Great Lakes region, New England, and the Maritime Provinces

BUTTERFLY HOST PLANT: Various blues, hairstreaks, metalmarks

Many species are available including both annuals and perennials; nearly all are excellent pollinator plants. Wild buckwheats are extremely important for several butterfly caterpillars. The flowers, in turn, are an important floral source for those same butterflies as mature adults. Wild buckwheat should be considered a "must-have" plant for pollinator gardens in California, the Great Basin, and the Southwest.

Wild geranium
(*Geranium* spp.)

EXPOSURE: Shade to part shade

SOIL MOISTURE: Average

BLOOM TIME: Spring

FLOWER COLOR: Variable

MAXIMUM HEIGHT: 3 feet (0.9 m)

REGIONS: Nationwide U.S. and Canada

Includes both perennial and annual species, most of which are visited by a variety of bee species including *Andrena*, *Augochlora*, and bumble bees. Common species include sticky purple geranium (*Geranium viscosissimum*) in the West, and Bicknell's cranesbill (*G. bicknellii*) and spotted geranium (*G. maculatum*) in the East. *Andrena distans* is an oligolectic bee of wild geranium.

Wild indigo (*Baptisia* spp.)

EXPOSURE: Sun

SOIL MOISTURE: Average

BLOOM TIME: Spring

FLOWER COLOR: Variable

MAXIMUM HEIGHT: 5 feet (1.5 m)

REGIONS: Central Plains east to Florida and north to New England and southern Ontario

BUTTERFLY HOST PLANT: Wild indigo duskywing, various skippers and sulfurs

Some species include blue wild indigo (*Baptisia australis*) and white wild indigo (*B. alba*). Visited primarily by long-tongued bees, especially queen bumble bees.

Wingstem (*Verbesina* spp.)

EXPOSURE: Shade to part sun

SOIL MOISTURE: Wet to average

BLOOM TIME: Summer

FLOWER COLOR: Yellow, white

MAXIMUM HEIGHT: 2 feet (0.6 m)

REGIONS: Midwestern, southeastern, and southwestern U.S.

Various wingstem species are widely distributed; the most common also tend to be the best nectar and pollen plants. Commercial seed sources are limited but increasing. Look for common wingstem (*Verbesina alternifolia*), golden crownbeard (*V. encelioides*), and white crownbeard (*V. virginica*).

Baccharis (*Baccharis* spp.)

EXPOSURE: Sun

SOIL MOISTURE: Dry

BLOOM TIME: Variable

FLOWER COLOR: White

MAXIMUM HEIGHT: 8 feet (2.4 m)

REGIONS: California, the Southwest, the Southeast, and Mid-Atlantic states

A large and diverse group of shrubs including spring-blooming mule-fat (*Baccharis salicifolia*), and coyotebrush (*B. pilularis*), which blooms in autumn.

Basswood (*Tilia americana*)

EXPOSURE: Sun to shade

SOIL MOISTURE: Average

BLOOM TIME: Early summer

FLOWER COLOR: White

MAXIMUM HEIGHT: 60 feet (18 m)

REGIONS: Eastern Canada and U.S. south to Texas and Florida

Basswood, and its nonnative cousin, littleleaf linden (*Tilia cordata*), are both noted nectar-producing trees among honey beekeepers. Other visitors include bumble bees, metallic green sweat bees, flies, and many wasp species.

Black locust
(*Robinia pseudoacacia*)

EXPOSURE: Sun

SOIL MOISTURE: Dry

BLOOM TIME: Late spring

FLOWER COLOR: White

MAXIMUM HEIGHT: 40 feet (12 m)

REGIONS: Nationwide U.S. and southern Canada

BUTTERFLY HOST PLANT: Silver-spotted skipper

Black locust is considered an excellent honey plant by many beekeepers, and it attracts a variety of other bees. The tree has sharp thorns on the leaf nodes. A shallow root system prone to suckering makes it hard to get rid of but also adaptable to harsh exposed sites like parking lot plantings. In some places it is considered an invasive weed because it self-sows readily.

Blueberry (*Vaccinium* spp.)

EXPOSURE: Sun

SOIL MOISTURE: Average

BLOOM TIME: Spring

FLOWER COLOR: White

MAXIMUM HEIGHT: 8 feet (2.4 m)

REGIONS: Northeast Canada south to Florida and east to the Great Lakes, also the Pacific Northwest

BUTTERFLY HOST PLANT: Striped hairstreak

Blueberries require evenly moist but well-drained soil that is quite acidic. The wild lowbush blueberry (*Vaccinium angustifolium*) is found throughout eastern Canada, the Great Lakes, and New England. Highbush blueberry (*Vaccinium corymbosum*) extends from New England east to Michigan; rabbiteye blueberry (*Vaccinium virgatum*) grows across the South from Florida to Texas and north to North Carolina. These and several other blueberry species are visited primarily by bees, especially Andrena mining bees, long-tongued bumble bees, and various mason bees. Honey bees and short-tongued bumble bees are known to rob nectar from blueberries by biting holes in the back of flowers.

Madrone (*Arbutus* spp.)

EXPOSURE: Sun to part shade
SOIL MOISTURE: Dry
BLOOM TIME: Mid-spring
FLOWER COLOR: White
MAXIMUM HEIGHT: 25 feet (7.6 m)
REGIONS: Pacific Northwest, California, Southwest
BUTTERFLY HOST PLANT: Various hair-streaks

Madrone is adapted to humid coastal sites as well as dry foothills and canyon areas. The most common species, Pacific madrone (*Arbutus menziesii*), forms clusters of white, bell-shaped flowers in mid-spring that develop into red berries, which provide food for birds, deer, and other wildlife. Madrone serves as a good erosion-control species and readily thrives in areas prone to frequent disturbance, including periodic burning.

Magnolia (*Magnolia* spp.)

EXPOSURE: Sun to shade
SOIL MOISTURE: Average
BLOOM TIME: Spring
FLOWER COLOR: White
MAXIMUM HEIGHT: 40 feet (12 m)
REGIONS: Southeastern U.S.

Magnolias are considered a very primitive flowering plant with a close association with various beetle pollinators, which were among their first evolutionary insect pollinators. Common species are also visited by bees, and they provide a notable ornamental addition to managed landscapes. Some popular species include the southern magnolia (*Magnolia grandiflora*), the bigleaf magnolia (*M. macrophylla*), and sweetbay (*M. virginiana*).

Manzanita
(*Arctostaphylos* spp.)

EXPOSURE: Sun

SOIL MOISTURE: Average

BLOOM TIME: Spring

FLOWER COLOR: White, pink

MAXIMUM HEIGHT: 20 feet (6 m)

REGIONS: California, Pacific Northwest

BUTTERFLY HOST PLANT: Various hairstreaks

Manzanitas comprise a huge group of flowering trees and shrubs, including the common or whiteleaf manzanita (*Arctostaphylos manzanita*), greenleaf manzanita (*A. patula*), bigberry manzanita (*A. glauca*), and the Pajaro manzanita (*A. pajaroensis*). The greatest diversity is found in California. Bumble bees and mason bees are common visitors to all of these. Kinnikinnick or bearberry (*A. uva-ursi*), a species found throughout most of the United States and Canada, also has excellent pollinator value and is a great groundcover.

Oceanspray
(*Holodiscus discolor*)

EXPOSURE: Sun to shade

SOIL MOISTURE: Average to dry

BLOOM TIME: Spring

FLOWER COLOR: White

MAXIMUM HEIGHT: 20 feet (6 m)

REGIONS: Western U.S. and Canada

BUTTERFLY HOST PLANT: Pale swallowtail

Ocean spray flowers provide pollen and nectar for not just bees, but also many other beneficial insects including wasps and syrphid flies. The thick foliage also provides cover for numerous birds, small mammals, and even tree frogs. Ocean spray is very adaptable and in the wild is found on dry rocky slopes, cool forest understories, and mountain peaks. It occurs most abundantly west of the Cascade Range near the Pacific coast.

Oregon grape
(*Mahonia* spp.)

EXPOSURE: Shade to part shade

SOIL MOISTURE: Dry

BLOOM TIME: Spring

FLOWER COLOR: Yellow

MAXIMUM HEIGHT: 12 feet (3.6 m)

REGIONS: Western U.S. and Canada

Also called hollyleaved barberry, this species exists in two widely separate ranges, an eastern population in the Great Lakes region of the United States and Canada and a western population in the Pacific Northwest. Bright yellow spring blossoms give way to clusters of blue fruit. Bees such as mining bees, mason bees, and bumble bees are the most common floral visitors. Handsome, evergreen foliage and multiseason interest make this a good landscaping shrub, except in windy sites.

Plum, cherry, almond, peach (*Prunus* spp.)

EXPOSURE: Sun to part shade

SOIL MOISTURE: Average

BLOOM TIME: Spring

FLOWER COLOR: Pink, white

MAXIMUM HEIGHT: 80 feet (24 m)

REGIONS: Nationwide U.S. and Canada

BUTTERFLY AND MOTH HOST PLANT: Red-spotted purple, eastern tiger swallowtail, various hairstreaks, various sphinx moths

The *Prunus* genus includes dozens of native and introduced spring-blooming trees and shrubs. Nearly all are high-value pollinator plants, supporting early-season bee species. In addition to cultivated species, numerous wild ones such as black cherry (*Prunus serotina*) are found across North America. *Andrena fenningeri* is a specialist bee of *Prunus* and *Salix* species.

Rabbitbrush
(*Chrysothamnus* spp.)

EXPOSURE: Sun

SOIL MOISTURE: Dry

BLOOM TIME: Fall

FLOWER COLOR: Yellow

MAXIMUM HEIGHT: 3 feet (0.9 m)

REGIONS: Western U.S. from Texas north to British Columbia

A common group of plants in arid regions of western North America, usually found growing with sagebrush. There are several different species, including yellow rabbitbrush (*Chrysothamnus viscidiflorus*) and Greene's rabbitbrush (*C. greenei*).

Redbud (*Cercis* spp.)

EXPOSURE: Sun to shade

SOIL MOISTURE: Average

BLOOM TIME: Spring

FLOWER COLOR: Magenta

MAXIMUM HEIGHT: 15 feet (4.5 m)

REGIONS: Eastern and southwestern U.S., southern Ontario

California redbud (*Cercis orbiculata*) is found in western states, while its eastern counterpart is the Eastern redbud (*Cercis canadensis*). Neither tree thrives at high altitudes or in extreme cold climates. Both attract primarily bee pollinators to the flowers in early spring. The leaves are sometimes used by leafcutter bees (genus: *Megachile*) for nesting material.

Rhododendron
(*Rhododendron* spp.)

EXPOSURE: Sun to shade

SOIL MOISTURE: Average

BLOOM TIME: Spring

FLOWER COLOR: Variable

MAXIMUM HEIGHT: 20 feet (6 m)

REGIONS: Eastern Canada and U.S. south to Texas; Pacific Northwest

Many species are found in the East including smooth azalea (*Rhododendron arborescens*), Piedmont rhododendron (*R. minus*), and early azalea (*R. prinophyllum*). Native western species include the Pacific rhododendron (*R. macrophyllum*) and the Cascade azalea (*R. albiflorum*). Numerous ornamental horticultural varieties are also available. Most species require acidic soil to thrive. Honey from some rhododendron plants has been reported as toxic to humans, but bees, especially bumble bees, readily visit the blossoms with no obvious ill effects.

Rose (*Rosa* spp.)

EXPOSURE: Sun to part shade

SOIL MOISTURE: Wet to dry

BLOOM TIME: Late spring to summer

FLOWER COLOR: White, yellow, red, pink

MAXIMUM HEIGHT: 12 feet (3.6 m)

REGIONS: Nationwide U.S. and Canada

Most roses need average to dry soil, but a few natives such as swamp rose tolerate wet roots. Open-faced flowers of native species are more attractive to pollinators than double-flowered ornamental varieties. Many roses produce little nectar, so pollen-collecting bees are the primary visitors. Common and adaptable species include Woods' rose (*Rosa woodsii*) and Nootka rose (*R. nutkana*) in the West; prairie rose (*R. arkansana*) in the Central Plains; and Virginia rose (*R. virginiana*), swamp rose (*R. palustris*), and Carolina rose (*R. carolina*) in the Midwest and East. *Eucera rosae* is a specialist bee of wild roses.

Serviceberry
(*Amelanchier* spp.)

EXPOSURE: Sun to part shade

SOIL MOISTURE: Average

BLOOM TIME: Spring

FLOWER COLOR: White

MAXIMUM HEIGHT: 15 feet (4.5 m)

REGIONS: Nationwide U.S. and Canada

BUTTERFLY HOST PLANT: Weidemeyer's admiral, western swallowtail

Also known as shadbush, several *Amelanchier* species are planted as ornamentals and as fruit trees. Many species are fire-dependent in the wild, growing most abundantly in forest areas recently cleared by burning. All are excellent pollinator plants, primarily attracting many bee species. Saskatoon serviceberry (*Amelanchier alnifolia*) is the most common western species, and common serviceberry (*A. arborea*) and Canadian serviceberry (*A. canadensis*) are most common in the East.

Sourwood
(*Oxydendrum arboreum*)

EXPOSURE: Sun to shade

SOIL MOISTURE: Average

BLOOM TIME: Early summer

FLOWER COLOR: White

MAXIMUM HEIGHT: 35 feet (10.6 m)

REGIONS: Mid-Atlantic and southeastern U.S.

Sourwood is well-known as a prolific nectar plant among honey beekeepers. These pyramid-shaped trees thrive in moist, peaty, acidic soil and produce striking red foliage in the fall. New ornamental cultivars such as "Chameleon" and "Mt. Charm" are becoming widely available.

Steeplebush, meadow-sweet (*Spiraea* spp.)

EXPOSURE: Sun

SOIL MOISTURE: Average to wet

BLOOM TIME: Late spring to early fall

FLOWER COLOR: White, pink, purple

MAXIMUM HEIGHT: 4 feet (1.2 m)

REGIONS: Nationwide Canada and U.S. except Florida and the desert Southwest

BUTTERFLY AND MOTH HOST PLANT: Various azures, Columbia silkmoth

Steeplebush (*Spiraea tomentosa*), and white meadowsweet (*S. alba*) are two widely distributed species with long bloom periods. Flower visitors include many butterflies, flies, and small bees.

Toyon (*Heteromeles arbutifolia*)

EXPOSURE: Sun

SOIL MOISTURE: Dry

BLOOM TIME: Spring

FLOWER COLOR: White

MAXIMUM HEIGHT: 8 feet (2.4 m)

REGIONS: California

Toyon once was a major component in California's chaparral ecosystem. The evergreen plant has dark green leathery leaves rising from multiple stems and long-lasting red berries that are consumed by birds. Toyon is ideal as a specimen shrub or as a screen when planted as a hedge. Bees are common flower visitors.

Tulip tree
(*Liriodendron tulipifera*)

EXPOSURE: Sun

SOIL MOISTURE: Average to wet

BLOOM TIME: Mid-spring

FLOWER COLOR: Yellow

MAXIMUM HEIGHT: 50 feet (15.2 m)

REGIONS: Eastern and southern U.S.

Tulip tree produces large amounts of nectar and is a major honey-producing tree in the southeastern United States. It is also visited by a number of native nectar-seeking pollinators. Tulip trees are considered to be an attractive addition to residential landscapes, and many cultivated varieties are now available from nurseries. When selecting this species, choose varieties that have bright yellow flowers, as those with greener color may be less attractive to pollinators.

Tupelo (*Nyssa* spp.)

EXPOSURE: Sun

SOIL MOISTURE: Average to wet

BLOOM TIME: Mid-spring

FLOWER COLOR: Green

MAXIMUM HEIGHT: 30 feet (9 m)

REGIONS: Eastern and southern U.S.

Both water tupelo (*Nyssa aquatica*) and the more widely distributed black gum (*N. sylvatica*) are well-known nectar plants among beekeepers, producing a very light, mild honey that commands a high market price. Florida beekeepers even place hives on floating platforms along river swamps to take advantage of the tupelo bloom. In addition to honey bees, tupelo supports a number of native pollinators including various bees and flies.

Wild lilac (*Ceanothus* spp.)

EXPOSURE: Sun to part shade

SOIL MOISTURE: Average

BLOOM TIME: Spring to summer

FLOWER COLOR: White, blue, pink

MAXIMUM HEIGHT: 8 feet (2.4 m)

REGIONS: Nationwide U.S. and southern Canada

BUTTERFLY HOST PLANT: Various species of blues, dusky wings, hairstreaks, and skippers

Recommended species include buckbrush (*Ceanothus cuneatus*) in western states, and New Jersey tea (*C. americanus*) in the Midwest and New England. Most species are slow-growing and difficult to establish from seed. Wild lilac is visited by an enormous diversity of bees and other pollinators such as syrphid and tachinid flies, mud daubers, spider wasps, sand wasps, and many butterflies.

Willow (*Salix* spp.)

EXPOSURE: Sun to part shade

SOIL MOISTURE: Wet to average

BLOOM TIME: Early spring

FLOWER COLOR: White, yellow

MAXIMUM HEIGHT: 30 feet (9 m)

REGIONS: Nationwide U.S. and Canada

BUTTERFLY AND MOTH HOST PLANT: Western tiger swallowtail, mourning cloak, Lorquin's and Weidemeyer's admirals, viceroy, red-spotted purple; various species of hairstreaks, skippers, and sphinx moths

Willows are an important spring pollen source for bees, often the first pollen available in many areas. For this reason, male plants are preferred; they are easily propagated by cuttings. Horticultural hybrids, including most weeping willows, are of little value. Pussy willows such as the native *Salix discolor* and the nonnative *S. caprea* and *S. cinerea* are among the better species. A number of specialist mining bees have been reported including *Andrena andrenoides*, *A. bisalicis*, *A. erythrogaster*, *A. fenningeri*, *A. illinoiensis*, *A. mariae*, *A. salictaria*, and *A. sigmundi*.

Yerba santa
(*Eriodictyon* spp.)

EXPOSURE: Sun

SOIL MOISTURE: Dry

BLOOM TIME: Spring

FLOWER COLOR: Purple, white

MAXIMUM HEIGHT: 8 feet (2.4 m)

REGIONS: Western U.S.

BUTTERFLY HOST PLANT: Lorquin's admiral

Most yerba santa species are limited to California with the notable exception of narrowleaf yerba santa (*Eriodictyon angustifolium*), whose range extends east to the Great Lakes. Bees and butterflies are common visitors, and honey bees produce a spicy, amber-colored honey from the nectar. Like other chaparral species, yerba santa thrives in recently burned areas.

Basil (*Ocimum* spp.)

EXPOSURE: Sun

SOIL MOISTURE: Average

BLOOM TIME: Summer

FLOWER COLOR: White

MAXIMUM HEIGHT: 2 feet (0.6 m)

REGIONS: Nationwide U.S. and southern Canada

Common sweet basil (*Ocimum basilicum*) is an easy-to-grow annual that supports bee visitors if allowed to bloom. Most basils are very susceptible to frost and cold-weather injury. In cold climates, start transplants indoors for earlier flowering plants.

Borage (*Borago officinalis*)

EXPOSURE: Sun

SOIL MOISTURE: Average

BLOOM TIME: Spring to summer

FLOWER COLOR: Blue

MAXIMUM HEIGHT: 2 feet (0.6 m)

REGIONS: Nationwide U.S. north to central Canada

An annual herb, occasionally grown as a field crop for oilseed. Stagger planting dates for longer bloom periods. Borage will adapt to many soil types but grows largest and flowers most prolifically in rich, fertile soils.

Catnip (*Nepeta* spp.)

EXPOSURE: Sun

SOIL MOISTURE: Average to dry

BLOOM TIME: Spring to summer

FLOWER COLOR: White, blue

MAXIMUM HEIGHT: 2 feet (0.6 m)

REGIONS: Nationwide U.S. and Canada

Wild types and ornamental species are available. Attractiveness to bees can vary depending on species and variety, but all are generally good bee plants.

Cosmos (*Cosmos bipinnatus*)

EXPOSURE: Sun

SOIL MOISTURE: Average

BLOOM TIME: Summer to autumn

FLOWER COLOR: White, pink, red

MAXIMUM HEIGHT: 6 feet (1.8 m)

REGIONS: Nationwide U.S. and southern
   Canada

A very common garden annual, visited by many bee species including green metallic sweat bees (*Agapostemon* spp.) and long-horned bees (*Mellisodes* spp.). Try mass plantings for maximum effect.

Lavender (*Lavandula* spp.)

EXPOSURE: Sun

SOIL MOISTURE: Dry

BLOOM TIME: Summer

FLOWER COLOR: Purple

MAXIMUM HEIGHT: 3 feet (0.9 m)

REGIONS: Moderate and warm climates
   across the U.S.

Both English lavender (*Lavandula angustifolia*) and French (*L. stoechas*) are productive bee plants, commonly attracting honey bees, mason bees, small carpenter bees, bumble bees, and both native and nonnative carder bees (*Anthidium* spp.).

Mexican sunflower
(*Tithonia rotundifolia*)

EXPOSURE: Sun

SOIL MOISTURE: Average to dry

BLOOM TIME: Summer

FLOWER COLOR: Orange, yellow

MAXIMUM HEIGHT: 6 feet (1.8 m)

REGIONS: Nationwide U.S.

A tall annual plant that thrives in warm climates. For maximum flower production, plant in full sun and remove faded blooms. In some areas, Mexican sunflower readily self-sows. Butterflies, bumble bees, and hummingbirds are some common floral visitors.

Mint (*Mentha* spp.)

EXPOSURE: Sun to shade

SOIL MOISTURE: Average to wet

BLOOM TIME: Summer

FLOWER COLOR: White, lavender

MAXIMUM HEIGHT: 1 foot (30 cm)

REGIONS: Nationwide U.S. and Canada

Recommended species include spearmint (*Mentha spicata*) and native wild mint (*M. arvensis*). Flies and wasps collect nectar from the flowers, and small bees are frequent flower visitors. Most mints prefer rich, wet soils, and many species spread by underground rhizomes.

Oregano (*Origanum* spp.)

EXPOSURE: Sun

SOIL MOISTURE: Dry

BLOOM TIME: Summer

FLOWER COLOR: White, pink

MAXIMUM HEIGHT: 1 foot (30 cm)

REGIONS: Nationwide U.S., southern Canada

Both sweet majoram (*Origanum majorana*) and oregano (*O. vulgare*) are attractive to bees, especially bumble bees.

Rosemary
(*Rosmarinus officinalis*)

EXPOSURE: Sun

SOIL MOISTURE: Dry

BLOOM TIME: Spring

FLOWER COLOR: Blue

MAXIMUM HEIGHT: 3 feet (0.9 m)

REGIONS: Warm regions of the U.S.

Rosemary is a small woody perennial with limited cold hardiness. It generally blooms in late winter to spring but sometimes repeats in autumn. Bees are the most common flower visitors. In cold climates, rosemary is sometimes grown in containers, then moved inside during the winter months and maintained as a house plant.

Russian sage
(*Perovskia atriplicifolia*)

EXPOSURE: Sun

SOIL MOISTURE: Dry

BLOOM TIME: Summer

FLOWER COLOR: Blue

MAXIMUM HEIGHT: 5 feet (1.5 m)

REGIONS: Nationwide U.S. and southern Canada

A very tough nonnative perennial plant with long-lasting blooms. Ideal for difficult sites, such as parking lot islands, and sunny building foundations. Bumble bees and carder bees are very common visitors.

Sweet clover (*Melilotus* spp.)

EXPOSURE: Sun

SOIL MOISTURE: Wet to dry

BLOOM TIME: Late spring to summer

FLOWER COLOR: White, yellow

MAXIMUM HEIGHT: 5 feet (1.5 m)

REGIONS: Nationwide U.S. and Canada

BUTTERFLY HOST PLANT: Various blues and sulfurs

Considered by many honey beekeepers to be one of the best nectar plants. Sweet clover attracts not only honey bees, but also a huge diversity of native bees. These weedy annual and biennial plants are very adaptable and easy to establish. Yellow sweet clover (*Melilotus officinalis*) is the most commonly available species.

Sweetvetch (*Hedysarum* spp.)

EXPOSURE: Sun

SOIL MOISTURE: Average to dry

BLOOM TIME: Late spring to early summer

FLOWER COLOR: Purple

MAXIMUM HEIGHT: 2 feet (0.6 m)

REGIONS: Western U.S. and Canada

Utah sweetvetch (*Hedysarum boreale*) is a native western species that has been extensively cultivated as a rangeland forage plant. "Timp" is a cultivar available from specialty seed producers. Utah sweetvetch produces prolific purple flower clusters and spreads by rhizomes, making it an excellent choice for dry western natural areas. Other less common sweetvetches are found throughout the western United States.

Vetch (*Vicia* spp.)

EXPOSURE: Sun to part shade
SOIL MOISTURE: Average
BLOOM TIME: Spring to summer
FLOWER COLOR: Purple, pink
MAXIMUM HEIGHT: 3 feet (0.9 m)
REGIONS: Nationwide U.S. and Canada
BUTTERFLY HOST PLANT: Various blues
  and sulfurs

Not to be mistaken with the invasive crownvetch (*Securigera varia*). Suitable species include the native perennial American vetch (*Vicia americana*) and the nonnative annual hairy or winter vetch (*Vicia villosa*), commonly used as a agricultural cover crop. All species attract a variety of bees especially bumble bees, honey bees, and leafcutter bees.

△ *Fiery skipper (*Hylephila phyleus)

caroliniana), checkermallow (*Sidalcea*), fanpetals (*Sida*), globemallow (*Sphaeralcea*), wild hollyhock (*Iliamna*)

Introduced. MALVACEAE: velvetleaf (*Abutilon theophrasti*), threelobe false mallow (*Malvastrum coromandelianum*), mallow (*Malva*)

Fiery skipper, *Hylephila phyleus*

Native. POACEAE: Bent grass (*Agrostis*), teal love grass (*Eragrostis hypnoides*), Kentucky bluegrass (*Poa pratensis*)

Introduced. POACEAE: Bermuda grass (*Cynodon dactylon*), bent grass (*Agrostis*)

Juba skipper, *Hesperia juba*

Native. POACEAE: slender hair grass (*Deschampsia elongata*), needle grass (*Stipa*), bluegrass (*Poa*)

Introduced. POACEAE: red brome (*Bromus rubens*)

Common branded skipper, *Hesperia comma*

Native. POACEAE: red fescue (*Festuca rubra*), bluegrass (*Poa*), muhly (*Muhlenbergia*), Thurber's needle grass (*Achnatherum thurberianum*), silver bluestem (*Bothriochloa saccharoides*), brome (*Bromus*), blue grama (*Bouteloua gracilis*)

Introduced. POACEAE: ryegrass (*Lolium*), brome (*Bromus*)

Peck's skipper, *Polites peckius*

Native. POACEAE: rice cutgrass (*Leersia oryzoides*), Kentucky bluegrass (*Poa pratensis*)

Tawny-edged skipper, *Polites themistocles*

Native. POACEAE: panic grass (*Panicum*), slender crabgrass (*Digitaria filiformis*), Kentucky bluegrass (*Poa pratensis*)

Crossline skipper, *Polites origenes*

Native. POACEAE: purpletop tridens (*Tridens flavus*), little bluestem (*Schizachyrium scoparium*)

Sachem skipper, *Atalopedes campestris*

Native. POACEAE: red fescue (*Festuca rubra*), St. Augustine grass (*Stenotaphrum secundatum*), hairy crabgrass (*Digitaria sanguinalis*)

△ *Sachem skipper (*Atalopedes campestris)

Introduced. POACEAE: Bermuda grass (*Cynodon dactylon*), Indian goosegrass (*Eleusine indica*)

Woodland skipper, *Ochlodes sylvanoides*

Native. POACEAE: canary grass (*Phalaris*), wild rye (*Elymus*), bluebunch wheatgrass (*Pseudoroegneria spicata*)

Introduced. POACEAE: Bermuda grass (*Cynodon dactylon*), bearded wheatgrass (*Elymus caninus*), colonial bent grass (*Agrostis capillaris*), wild oat (*Avena fatua*)

△ *Woodland skipper* (Ochlodes sylvanoides)

Delaware skipper, *Anatrytone logan*

Native. POACEAE: big bluestem (*Andropogon gerardii*), silver plume grass (*Saccharum alopecuroides*), switch grass (*Panicum virgatum*)

Swallowtails (Papilionidae)

Clodius Parnassian, *Parnassius clodius*

Native. FUMARIACEAE: bleeding heart (*Dicentra*)

Pipevine swallowtail, *Battus philenor*

Native. ARISTOLOCHIACEAE: Dutchman's pipe (*Aristolochia*). CONVOLVULACEAE: tropical white morning glory (*Ipomoea alba*). POLYGONACEAE: knotweed (*Polygonum*)

Introduced. POLYGONACEAE: black bindweed (*Polygonum convolvulus*)

Black swallowtail, *Papilio polyxenes*

Native. APIACEAE: angelica (*Angelica*), American wild carrot (*Daucus pusillus*), Canadian honewort (*Cryptotaenia canadensis*), Scottish licorice-root (*Ligusticum scoticum*), water hemlock (*Cicuta*), zizia (*Zizia*), umbrellawort (*Tauschia*). RUTACEAE: desert rue (*Thamnosma*)

Introduced. APIACEAE: dill (*Anethum graveolens*), Queen Anne's lace (*Daucus carota*), fennel (*Foeniculum*), wild celery (*Apium graveolens*), poison hemlock (*Conium maculatum*), garden lovage (*Levisticum officinale*). RUTACEAE: rue (*Ruta graveolens*), gasplant (*Dictamnus albus*)

△ *Black swallowtail* (Papilio polyxenes)

Anise swallowtail, *Papilio zelicaon*

Native. APIACEAE: angelica (*Angelica*), American wild carrot (*Daucus pusillus*), common cowparsnip (*Heracleum*

△ *Painted lady* (Vanessa cardui)

Atlantis fritillary, *Speyeria atlantis*

Native. VIOLACEAE: violet (*Viola*)

American lady, *Vanessa virginiensis*

Native. ASTERACEAE: cudweed (*Gnaphalium, Pseudognaphalium*), pussytoes (*Antennaria*), pearly everlasting (*Anaphalis margaritacea*), sagebrush (*Artemisia*), sunflower (*Helianthus*). FABACEAE: lupine (*Lupinus*). URTICACEAE: nettle (*Urtica*)

Introduced. ASTERACEAE: Canada thistle (*Cirsium arvense*), plumeless thistle (*Carduus*), dusty miller (*Senecio cineraria*). MALVACEAE: mallow (*Malva*), hollyhock (*Alcea rosea*). URTICACEAE: nettle (*Urtica*). SCROPHULARIACEAE: snapdragon (*Antirrhinum*)

Painted lady, *Vanessa cardui*

Native. ASTERACEAE: thistle (*Cirsium*), sagebrush (*Artemisia*), common yarrow (*Achillea millefolium*), pearly everlasting (*Anaphalis margaritacea*), sunflower (*Helianthus*), helianthella (*Helianthella*). MALVACEAE: alkali mallow (*Malvella leprosa*), desert globemallow (*Sphaeralcea ambigua*). CHENOPODIACEAE: lambsquarters (*Chenopodium album*). FABACEAE: lupine (*Lupinus*), clover (*Trifolium*). URTICACEAE: stinging nettle (*Urtica dioica*). LABIATAE: mint (*Mentha*), sage (*Salvia*), betony (*Stachys coccinea*)

Introduced. ASTERACEAE: thistle (*Cirsium*), globe artichoke (*Cynara scolymus*), Scotch thistle (*Onopordum acanthium*), milk thistle (*Silybum marianum*), plumeless thistle (*Carduus*), burdock (*Arctium*), blessed thistle (*Cnicus benedictus*), pot marigold (*Calendula officinalis*), knapweed (*Centaurea*), dusty miller (*Senecio cineraria*). BORAGINACEAE: borage (*Borago officinalis*), viper's bugloss (*Echium vulgare*), comfrey (*Symphytum officinale*). MALVACEAE: mallow (*Malva*), hollyhock (*Alcea rosea*). CHENOPODIACEAE: common beet (*Beta vulgaris*), lambsquarters (*Chenopodium album*). FABACEAE: alfalfa (*Medicago sativa*), garden pea (*Pisum sativum*), clover (*Trifolium*). URTICACEAE: dwarf nettle (*Urtica urens*). SOLANACEAE: petunia (*Petunia*). CONVOLVULACEAE: sweet potato (*Ipomoea batatas*)

△ *Red admiral* (Vanessa atalanta)

West Coast lady, *Vanessa annabella*

Native. MALVACEAE: island mallow (*Lavatera assurgentiflora*), Mendocino bushmallow (*Malacothamnus fasciculatus*), alkali mallow (*Malvella leprosa*), checkerbloom (*Sidalcea*), desert globemallow (*Sphaeralcea ambigua*). URTICACEAE: stinging nettle (*Urtica dioica*)

Introduced. MALVACEAE: mallow (*Malva*), hollyhock (*Alcea rosea*). URTICACEAE: dwarf nettle (*Urtica urens*)

Red admiral, *Vanessa atalanta*

Native. URTICACEAE: nettle (*Urtica*), mother of thousands (*Soleirolia soleirolii*), pellitory (*Parietaria*), small-spike false nettle (*Boehmeria cylindrica*). CANNABACEAE: common hop (*Humulus lupulus*)

Introduced. URTICACEAE: nettle (*Urtica*)

△ *Milbert's tortoiseshell* (Aglais milberti)

Milbert's tortoiseshell, *Aglais milberti*

Native. URTICACEAE: stinging nettle (*Urtica dioica*)
Introduced. URTICACEAE: stinging nettle (*Urtica dioica*)

Mourning cloak, *Nymphalis antiopa*

Native. SALICACEAE: willow (*Salix*), cottonwood (*Populus*). BETULACEAE: birch (*Betula*), alder (*Alnus*). ACERACEAE: maple (*Acer*). ULMACEAE: elm (*Ulmus*), hackberry (*Celtis*). CANNABACEAE: common hop (*Humulus lupulus*), red mulberry (*Morus rubra*). OLEACEAE: white ash (*Fraxinus americana*). ROSACEAE: mountain ash (*Sorbus*), white meadowsweet (*Spiraea alba*), blackberry (*Rubus*), rose (*Rosa*). TILIACEAE: basswood (*Tilia americana*)

Introduced. ROSACEAE: common pear (*Pyrus communis*)

Question mark, *Polygonia interrogationis*

Native. ULMACEAE: elm (*Ulmus*), hackberry (*Celtis*). CANNABACEAE: common hop (*Humulus lupulus*). URTICACEAE: stinging nettle (*Urtica dioica*), small-spike false nettle (*Boehmeria cylindrica*)

Introduced. CANNABACEAE: Japanese hop (*Humulus japonicus*). URTICACEAE: stinging nettle (*Urtica dioica*)

△ *Question mark* (Polygonia interrogationis)

△ *Baltimore checkerspot*
(Euphydryas phaeton*)*

Eastern comma, *Polygonia comma*

Native. ULMACEAE: American elm (*Ulmus americana*). CANNABACEAE: common hop (*Humulus lupulus*). URTICACEAE: stinging nettle (*Urtica dioica*), Canadian woodnettle (*Laportea canadensis*), small-spike false nettle (*Boehmeria cylindrica*)

Introduced. URTICACEAE: stinging nettle (*Urtica dioica*)

Satyr comma, *Polygonia satyrus*

Native. URTICACEAE: stinging nettle (*Urtica dioica*). CANNABACEAE: common hop (*Humulus lupulus*)

Common buckeye, *Junonia coenia*

Native. PLANTAGINACEAE: plantain (*Plantago*). SCROPHULARIACEAE: owl's-clover (*Orthocarpus*), false foxglove (*Agalinis*), roving sailor (*Maurandella antirrhiniflora*), yaupon blacksenna (*Seymeria cassioides*), American blue-hearts (*Buchnera americana*), speedwell (*Veronica*). VERBENACEAE: fogfruit (*Phyla*)

Introduced. PLANTAGINACEAE: plantain (*Plantago*). SCROPHULARIACEAE: toadflax (*Linaria*), garden snapdragon (*Antirrhinum majus*), Kenilworth ivy (*Cymbalaria muralis*)

Baltimore checkerspot, *Euphydryas phaeton*

Native. SCHROPHULARIACEAE: white turtlehead (*Chelone glabra*), false foxglove (*Aureolaria*), Allegheny monkeyflower (*Mimulus ringens*), Canadian lousewort (*Pedicularis canadensis*), hairy beardtongue (*Penstemon hirsutus*), carpenter's square (*Scrophularia marilandica*), speedwell (*Veronica*). CAPRIFOLIACEAE: honeysuckle (*Lonicera*), coralberry (*Symphoricarpos orbiculatus*), arrowwood (*Viburnum dentatum*). OLEACEAE: white ash (*Fraxinus americana*)

Introduced. CAPRIFOLIACEAE: dwarf honeysuckle (*Lonicera xylosteum*)

Chalcedona checkerspot, *Euphydryas chalcedona*

Native. SCHROPHULARIACEAE: Indian paintbrush (*Castilleja*), blue-eyed Mary (*Collinsia*), bush monkeyflower (*Diplacus*), beardtongue (*Penstemon*), California figwort (*Scrophularia californica*), lousewort (*Pedicularis*). VALERIANACEAE: seablush (*Plectritis*). CAPRIFOLIACEAE: honeysuckle (*Lonicera*), snowberry (*Symphoricarpos*)

△ *Gorgone checkerspot*
(Chlosyne gorgone*)*

Silvery checkerspot, *Chlosyne nycteis*

Native. ASTERACEAE: cutleaf coneflower (*Rudbeckia laciniata*), crownbeard (*Verbesina*), aster (*Symphyotrichum*), whitetop (*Doellingeria*), sunflower (*Helianthus*), Canadian horseweed (*Conyza canadensis*), goldenrod (*Solidago*)

Gorgone checkerspot, *Chlosyne gorgone*

Native. ASTERACEAE: sunflower (*Helianthus*), great ragweed (*Ambrosia trifida*), showy goldeneye (*Heliomeris multiflora*)

Introduced. ASTERACEAE: giant sumpweed (*Cyclachaena xanthifolia*)

△ *Field crescent* (Phyciodes pulchella*)*

Mylitta crescent, *Phyciodes mylitta*

Native. ASTERACEAE: thistle (*Cirsium*). SCROPHULARIACEAE: seep monkeyflower (*Mimulus guttatus*)

Introduced. ASTERACEAE: thistle (*Cirsium*), milk thistle (*Silybum marianum*), Italian plumeless thistle (*Carduus pycnocephalus*)

Pearly crescent, *Phyciodes tharos*

Native. ASTERACEAE: aster (*Oclemena, Symphyotrichum*), whitetop aster (*Sericocarpus*)

Field crescent, *Phyciodes pulchella*

Native. ASTERACEAE: aster (*Eurybia, Eucephalus, Symphyotrichum*), Bigelow's tansyaster (*Machaeranthera bigelovii*)

Common ringlet, *Coenonympha tullia*

Native. POACEAE: needlegrass (*Stipa*), Kentucky bluegrass (*Poa pratensis*)

Little wood-satyr, *Megisto cymela*

Introduced. POACEAE: orchard grass (*Dactylis glomerata*)

Common wood-nymph, *Cercyonis pegala*

Native. POACEAE: bluestem (*Andropogon*), purpletop tridens (*Tridens flavus*), porcupinegrass (*Hesperostipa spartea*)

Introduced. POACEAE: wild oat (*Avena fatua*)

△ *Common wood nymph* (Cercyonis pegala)

PHOTOGRAPHY CREDITS

© Toby Alexander/USDA–NRCS 178 b, 182, 300 r, 301 l

© Jennifer Anderson @ USDA–NRCS PLANTS Database181 b

© John Ascher 147, 250

© Gene Barickman 285 b

© Robert Behrstock 168 tr & br, 300 l, 303 r, 316 tl, 317

© Matt Below 290 tl

© Tom Bentley 338 t

© Betsy Betros 226, 247

© Scott Black/The Xerces Society 79, 157 b

© William Bouton 63, 65 t & m, 69 t, 324–327, 328 t, 329, 330, 331 t, 332–336, 337 b, 339 b

© Lynda Boyer 112–114, 122, 126 b

© Guy Bruyea 191

© Charles T. Bryson, USDA–ARS, Bugwood.org 311 r

© Stephen L. Buchmann 26

© James L. Cane 28, 48 t, 11, 174 r

© David Cappaert, Michigan State University, Bugwood.org 37, 222 b

© Allen Casey/USDA–NRCS 321 l

© Rollin Coville ii, vi, ix, x, 4, 15, 19 m, 20 b, 24 b, 27, 30, 34 b, 38, 41, 43, 47, 50, 52, 53 tr, 69 b, 80 b, 101, 136, 213 tl & bl, 214, 221, 223 t, 224 t & bl, 229, 236–238, 240, 242, 243, 245, 246, 248, 249, 251, 252, 255, 256, 258, 260, 355

© Julie Craves 244

© Lloyd Crim 142 b

© Rob Curtis/The Early Birder 231, 234

© Kim Davis & Mike Stangeland 328 b, 337 t, 338 b

© Dan Dinelli 196

© Chris Evans, River to River CWMA, Bugwood.org 302 l

© Elise Fog 22, 36, 44, 53 tl, 223 b, 224 br, 340–341

© Alistair Fraser 10 br

© Hannah Gaines 99, 213 ml, 292 r

© Rod Gilbert 7, 9, 18, 23, 57 t, 59 b, 93, 94, 104 t, 105, 153, 166, 167, 168 bl, 213 ml lower, 222 t, 284, 285 tr, 288 r, 290 tr, 291 tr, 297 l, 298, 301 r, 304 l, 305 l, 306, 308, 309 l, 310 l, 316 tr, 322 r

© Sarah Greenleaf 102

© M.J. Hatfield 164 l, 239, 287 l, 292 l, 294 l, 299 l

© Kirk Henderson 84

© Jennifer Hopwood/The Xerces Society163

© iStockphoto.com/Noam Armonn 35 b

© iStockphoto.com/Cynthia Baldauf 82 b

© iStockphoto.com/Brandon Blinkerberg 60 b

© iStockphoto.com/Pamela Burley 24 t

© iStockphoto.com/Tony Campbell 8 br

© iStockphoto.com/Markus Divis v

© iStockphoto.com/Steve Geer 189

© iStockphoto.com/hsvrs 169 bl

© iStockphoto.com/David Huss 74 t

© iStockphoto.com/Julie Kendall 314 l

© iStockphoto.com/Rich Legg 288 l

© iStockphoto.com/Frank Leung 20 t

© iStockphoto.com/Dawn Nichols 8 bl

© iStockphoto.com/Simon Phipps 155

© iStockphoto.com/Ryan Poling 66

© iStockphoto.com/Hanson Quan 159

© iStockphoto.com/Piotr Skubisz 319 r

© iStockphoto.com/Olga Utlyakova xi

© David Inouye 19 bl, 21, 56 t, 61, 259

© Johanna James–Heinz i, 78

© Paul Jepson 86

© JFNew 3 b, 97, 98, 282

© Preston Scott Justis 59 t

© Neal Kramer 286 tr, 289 r, 293 r, 295 tl, 296 r, 297 r, 302 r, 310 r, 313

© H. J. Larson, Bugwood.org 132

© Eric Mader/The Xerces Society 17 b, 55 b, 67, 90 b, 95, 121 cl, 130, 134, 142 t, 149, 150, 161, 168 tc & bc, 170 b, 171 b, 285 tl, 291 tl, 293 l, 295 b, 316 b, 318 l, 319 l

© Kevin Matteson 160

© Gary McDonald 219, 228

© Robert H Mohlenbrock @ USDA–NRCS PLANTS Database169 br, 303 l

© Beatriz Moisset 241

© Lynn Monroe 257

© Bruce Newhouse/Salix Associates 220, 235

© David Phipps 193

© Bryan R. Reynolds 19 br, 54, 55 t, 56 b, 57 b, 60 t & m, 64, 65 b, 68, 70, 83, 213 tr

© Lief Richardson 10 bl

© Dana Ross 57 m

© Edward S. Ross 35t, 42, 119, 121 l, 152, 156, 227 l, 232, 233

© Lynette Schimming 218

© Peter C. Schroeder 11

© William Settle 173

© Alan Shadow/USDA–NRCS 121 r, 281

© Matthew Shepherd/The Xerces Society 3, 6, 8 t, 19 t, 74 b, 75, 80 t, 87 t, 88, 92, 96 b, 100, 106, 110, 115, 116, 137, 140, 148, 162, 165 l, 170 t, 171 t, 174 l, 212, 213 mr, 230, 262, 286 tl, 290 b, 296 l, 315 b, 318 r

© Robert Sivinski 295 tr

© David Smith, delawarewildflowers.org 17 t, 168 tl, 179, 181 t, 186, 188, 194, 199, 200, 204, 207, 286 b, 307 r, 311 l, 315 tl, 320 r, 322 l

© Christine Taliga 90 m, 121cr, 164 r, 165 r, 287 r, 289 l, 291 b

© Ann Thering 331 b, 339 t

© Rebecca Tonietto 201

© Mark Turner 169 tl, 190, 299 r, 305 r, 314 r, 321 r, 333

© Katharina Ullmann/The Xerces Society 87 b

© Maria Ulrice/Iowa Living Roadsides Trust 202

USDA–ARS 82 t, 129, 213 bc

USDA–ARS/Stephen Ausmus 73

USDA–ARS/Scott Bauer 39, 123

USDA–ARS/Peggy Greb 5, 77

USDA–ARS/Edward McCain 169 tc

USDA–ARS/Keith Walker 185

USDA–NRCS PLANTS Database/Bureau of Land Management 307 l

USDA–NRCS/Irv Cole 126 t

USDA–NRCS/Brian Miller 124

USDA–NRCS/Jeff Vanuga 127

© Wendy VanDyk Evans, Bugwood.org 309 r

© Mace Vaughan/ The Xerces Society 2, 16, 33, 34 t, 90 t, 91, 96 t, 103, 104 b, 117, 135, 138, 157 t, 169 tr & bc, 175–177, 178 t, 211, 213 br, 253, 254, 294 r, 312, 315 tr, 320 l

© Marcelino Vilaubi xii, 304 r

© Keith Webber, Jr. 12

© Steve Werblow, Deere & Co, 10 t

© Alex Wild 48 b, 51, 53 b, 71, 227 r

OTHER .... LL ENJOY

The Backyard F

An information _____ species of birds.
320 pages. Pape

The Backyard F

A complete gui _____ ble anywhere —
from one's own
368 pages. Pape

The Bird Watch

Hundreds of qu _____ of Ornithology.
400 pages. Flexi

The Family But

Projects, activit _____ rican species.
176 pages. Paper. ISBN 978-1-58017-292-9.
Hardcover. ISBN 978-1-58017-335-3.

Keeping a Nature Journal, by Clare Walker Leslie & Charles E. Roth.
Simple methods for capturing the living beauty of each season.
224 pages. Paper with flaps. ISBN 978-1-58017-493-0.

The Life Cycles of Butterflies, by Judy Burris & Wayne Richards.
A visual guide, rich in photographs, showing 23 common backyard butterflies from
egg to maturity. Winner of the 2007 Teachers' Choice Children's Book Award!
160 pages. Paper. ISBN 978-1-58017-617-0.
Hardcover with jacket. ISBN 978-1-58017-618-7.

These and other books from Storey Publishing are available
wherever quality books are sold or by calling 1-800-441-5700.

Visit us at www.storey.com.